STUDIES IN HARMONIC ANALYSIS

Studies in Mathematics

The Mathematical Association of America

Studies in Mathematics

Volume 13

STUDIES IN HARMONIC ANALYSIS

J. M. Ash, editor

DePaul University

Published and distributed by

The Mathematical Association of America

Library of Congress Catalog Card Number 76-16431

Complete Set ISBN 0-88385-100-8
Vol. 13 ISBN 0-88385-113-X

Printed in the United States of America

Current printing (last digit):

10 9 8 7 6 5 4 3 2 1

ACKNOWLEDGMENTS

This volume constitutes the proceedings of the expository conference, "A Survey of Harmonic Analysis", which was held at DePaul University in Chicago on June 29–July 2, 1974, in celebration of the University's seventy-fifth anniversary. The help of Jerry Goldman, Karen Kenny, and Jim Nuccio was indispensable in making arrangements for the 11 speakers and over 200 conferees. Karen Kenny's typing was another *sine qua non* for this book. I would like to thank DePaul University and Rev. W. T. Cortelyou, the University of Chicago, Cotter & Company, and an anonymous DePaul alumnus for their contributions which made the conference possible. Finally, Guido Weiss's expert advice is gratefully acknowledged.

J. MARSHALL ASH

INTRODUCTION

For more than a century harmonic analysis has been one of the major branches of mathematics. For many years it served as a meeting ground for real and complex analysis.

Harmonic analysis has in recent years remained one of the most popular and rapidly expanding areas of mathematics. Three forces contributing to this growth are (1) the revitalization of old areas such as the study of Fourier series and multiple trigonometric series, (2) the discovery and development of generalizations, from series or integrals defined on Euclidean spaces, to more general Fourier transforms consisting of matrices, or even operators on Hilbert spaces, and (3) the application of methods from seemingly far removed disciplines such as probability theory, topological group theory, and partial differential equations.

Since harmonic analysis is so big and touches so many other fields, the most a volume such as this can hope to do is give the flavor of the field. It cannot even represent every major branch of harmonic analysis; for example, we regret the absence of sections on partial differential equations and on number theory.

For a general overview preliminary to reading this book, the reader might profit from looking at the article,"Harmonic Analysis," by Guido Weiss in *Studies in Real and Complex Analysis*—the third volume of this series.

A rough categorization of the contents of this volume can be made by saying that the articles of Zygmund, Hunt, Fefferman, Ash, Stein, and Burkholder deal with classical topics; and those of Meyer, Graham, Weiss, Sally, and Vági are concerned with branches of more recent growth.

The articles of Stein, Zygmund, and Fefferman were transcribed from videotape.

Antoni Zygmund's article starts out with a discussion of some highlights of nineteenth and early twentieth century work in Fourier series. He then considers in somewhat more detail the major developments in the theory of trigonometric series over the past half century. This latter discussion is enhanced by the fact of his own central position in the field throughout this period. Since Carleson's theorem on the convergence of Fourier series is discussed in detail by R. Hunt in the following article, it is omitted here.

One measure of Zygmund's influence is that among the ten other authors represented in this volume, three are his students and three more are students of his students. This is a good place to point out that his book, *Trigonometric Series* (second corrected edition published by Cambridge University Press in 1968) contains a vast amount ot material fundamental to harmonic analysis. Almost everything mentioned in our book either can be found in *Trigonometric Series*, generalizes something there, or at least grows out of something there.

Richard Hunt takes a novel approach to the exposition of Carleson's theorem. This theorem states that the Fourier series of an $L^2(0, 2\pi)$ function converges almost everywhere. Since Kolmogorov's 1926 example of an L^1 $(0, 2\pi)$ function with everywhere divergent Fourier series had encouraged a widespread belief in a similar fate for some L^2 function, Carleson's theorem electrified the mathematical community. I vividly remember an enthusiastic fellow graduate student bursting into a seminar room with the news. Unfortunately its proof remains technically difficult. Hunt's idea is to give the flavor of the subject by leading the reader through theorems upon which Carleson's theorem is based. Whenever these theorems, in turn, become too technical, he provides simplified versions. One has to pick up pencil and paper and work here, but we strongly urge the novice to do so. The effort will be amply rewarded. Incidentally, Hunt himself proved that $f \in$

$L^p, p > 1$, is also sufficient for convergence a.e.; his reference [10] contains this and is reasonably readable. Many people refer to this result as the Carleson-Hunt theorem.

Charles Fefferman starts by stating and sketching the proof of some standard theorems concerning H^p functions on the unit disc. He then passes to some theorems about $H^p(R^n)$ that have also been known for quite some time. One of these, A. P. Calderón's theorem concerning the existence of boundary values for non-tangentially bounded harmonic functions, is discussed at length and leads into a very clear discussion of martingales. The author, who proved the dual of H^1 to be represented by all functions of bounded mean oscillation, gives a very intuitive and lucid discussion of H^1 and its dual via examples. Finally, several modern definitions of H^p are given which show the Poisson kernel to be only an artifice, thus helping "free" H^p and hence to some extent all of harmonic analysis from its dependence upon complex function theory.

Marshall Ash's article on multiple trigonometric series begins with a very elementary discussion of numerical double series which shows the difficulty of finding a single suitable definition of convergence for such series. A number of classical facts about trigonometric series (concerning convergence and divergence, localization, uniqueness, and so forth) are stated and the degree to which each one can be extended to higher dimensions is considered. Emphasis is placed on the rectangular methods of convergence rather than the very important circular method (which is difficult to work with).

The main theme of Elias Stein's article is to describe first some very old and well-known inequalities from the theory of Fourier transforms on Euclidean n-space, and then go on to consider some facet of each inequality which had until very recently been overlooked. (For example, the constant 1 in the Hausdorff-Young inequality for Fourier integrals can be improved, while this is not the case for Fourier series.) Stein examines several convolution operators from the standpoint of the symmetries they preserve. This leads him to non-isotropic dilations, the Heisenberg group, and other interesting topics.

Donald Burkholder's thesis is that the Brownian motion point of view is quite helpful in proving theorems of harmonic analysis on $[0, 2\pi)$ or on R^n. To illustrate this method (which has become rather popular in the last few years), Burkholder presents Burgess Davis's new probabilistic proof of Kolmogorov's classical estimate for the size of the Hilbert transform of an $L^1(0, 2\pi)$ function. As an added bonus, Davis's proof discovers a hitherto unknown best constant which cannot, at present, be obtained by any other method. (The constant involves the quantity $\pi^{-2}(1 - 1/3^2 + 1/5^2 - \dots)$ which has a long mathematical history. Perhaps this will add a new impetus for trying to answer the old question of whether this quantity is rational.)

Yves Meyer starts with a simple difference equation and a simple differential equation. From the common elements of their solutions he distills a very general problem—the problem of solving $S*f = 0$, where S is a known distribution and f is an unknown function on R or R^n with a certain amount of postulated smoothness. He gives the reader insight into the beautiful theory that has arisen in the solution of various aspects of this problem.

Colin Graham's article is divided into two distinct parts. The first is an elementary introduction to the general theory of locally compact abelian groups. It covers much material quickly and clearly. The second part deals with certain "thin" sets—sets of spectral synthesis, interpolation sets, Kronecker sets, Helson sets, etc.,—that have provided an important area of research in the last decades. Research in this area has been motivated by the attempt to classify sets of uniqueness and sets of multiplicity in the theory of Fourier series of one variable. A most interesting point is the appearance of the non-classical groups (discussed in the first part) in the proofs of theorems concerning the interval $[0, 2\pi)$.

Guido Weiss gives a general discussion of harmonic analysis on compact groups. His article contains many basic facts upon which both Paul Sally and Colin Graham draw. First he determines the structure of compact groups via the Peter–Weyl theorem. He then does some harmonic analysis on a particular compact group $SU(2)$—the special unitary group. There are good reasons for studying this group. Although $SU(2)$ is the "first" and simplest non-abelian compact topological group, it nevertheless suggests what can be done on the class of compact Lie groups with only a slight loss of generality, and so it is a good testing ground for conjectures.

Paul Sally was faced with a difficult problem: How do you explain enough about the underlying locally compact spaces and about group representations, and still have time left to consider doing harmonic analysis? His solution involved (1) taking a fair amount of background as prerequisite, (2) stating a lot of theorems without proof, and (3) working out some examples in detail. The last requires an effort with pencil and paper from the novice reader that becomes increasingly greater as the article progresses, but which is well worth the effort. In this way he is able to come to grips with the harmonic analysis (of locally compact groups) itself in sections five and six. Again the method is to prove things for the special cases, while only stating the general results (that are often very difficult to prove); yet, in some instances, the special proofs are just as informative.

Stephen Vági considers the question, "What are the natural generalizations of the unit disc in higher dimensions?" He traces the history of this problem from 1907 (when Poincaré discovered that the "obvious" generalization of the Riemann mapping theorem fails). In the second part of his talk he briefly surveys some of the harmonic analysis in several variables related to Cartan and Siegel domains. These are the domains which provide an acceptable answer to the above question. Vági's article draws on many subfields of harmonic analysis so the novice reader would do well to look at several other articles in this book before turning to this one.

As was mentioned before in the Acknowledgments, these expositions were originally given in a conference held at DePaul University in Chicago on June 29–July 2, 1974. Each of the talks has been recorded by the aid of videotape. The articles of Stein, Fefferman, and Zygmund are rather faithful transcriptions of the recordings of the actual lectures they gave.

J. MARSHALL ASH

CONTENTS

NOTES ON THE HISTORY OF FOURIER SERIES*

Antoni Zygmund

1. INTRODUCTION

I plan to discuss some developments of trigonometric series in the first half of the twentieth century, but initially I shall have to go back to the nineteenth century. (This may be taken as a fringe benefit.) I will begin with elementary notions.

By definition, a trigonometric series is an infinite linear combination of exponentials:

$$\sum_{n=-\infty}^{\infty} c_n e^{inx}.$$

If this series represents anything, it represents a function $f(x)$ of

*This talk was transcribed (with the aid of videotape) from the hour lecture.

period 2π:

$$f(x + 2\pi) = f(x).$$

The study of trigonometric series is undertaken for several reasons; one of them is that this is a topic important and interesting in itself; another is that the methods applied here can be applied to many other problems. Thus trigonometric series are the central topic of these remarks but their study has many ramifications. Some of these extensions are relatively easy, others are more difficult. Relatively easy, for example, are extensions from trigonometric series to trigonometric integrals where summation is replaced by integration with respect to a continuous variable. Other developments are more difficult. For example, I shall not say anything about multiple trigonometric series. Certain results are automatic in passing from one to several variables; other things are more difficult. As a matter of fact, many of the most interesting problems nowadays are associated with several variables. For the time being the topic of one variable has been to a considerable degree exhausted.

Suppose you represent a certain function $f(x)$, where $f(x + 2\pi) = f(x)$, by a trigonometric series:

$$f(x) = \sum_{n=-\infty}^{\infty} c_n e^{inx}. \tag{1}$$

Having this representation one may formally determine the coefficients c_m by multiplying the equation by e^{-imx} and integrating over any interval of length 2π, e.g., $(0, 2\pi)$:

$$c_m = c_m(f) = \frac{1}{2\pi} \int_0^{2\pi} f(x) e^{-imx}\, dx. \tag{2}$$

Conversely, starting with a function $f(x)$ of period 2π, one may define its *Fourier coefficients*—$c_n = c_n(f)$—by equation (2) and then form the series (1) which by definition is the Fourier series of the function $f(x)$.

The main problem here is as follows: given the function f, determining the $c_n(f)$ like this and having formed the Fourier series, under what conditions does the series represent the function f? This is the central problem of the theory. The formal beginnings are associated with the names of Euler and Fourier, but of course the topic had a prehistory in connection with partial differential equations, which I won't discuss here. I mentioned that Euler and Fourier are responsible for the beginnings of the theory, primarily formal things. But now in the 19th century we have important developments and big names. Let me mention three names which represent considerable developments in this period. The first is Dirichlet, the second Riemann, and the third Cantor. These are the central names of the theory in the nineteenth century.

What is the main contribution of Dirichlet? He was the first mathematician to investigate the validity of the representation of the function by its Fourier series. He considered, of course, only the convergence of series. In his memoir—it was around 1837—he showed that if the function has a very simple structure, essentially if its graph can be split into a finite number of monotone curves, then the series does converge and does represent the function f. Concretely,

$$S_n(x) = \sum_{k=-n}^{+n} c_k(f)e^{ikx}$$

tends at each point to the mean value of the limits of f from the left and right (see Figure 1).

This is the so-called Dirichlet test for the convergence of Fourier series. Sometimes it is called the Dirichlet-Jordan test, because a theorem of Jordan states that a function of bounded variation is the difference of two monotone functions, and so automatically has a Fourier series convergent everywhere. But essentially this result is already in Dirichlet. Historically, this was the first test for convergence. Later there was a flood of papers on the subject—especially around the turn of the century. Some of the results are important, but this result was basic.

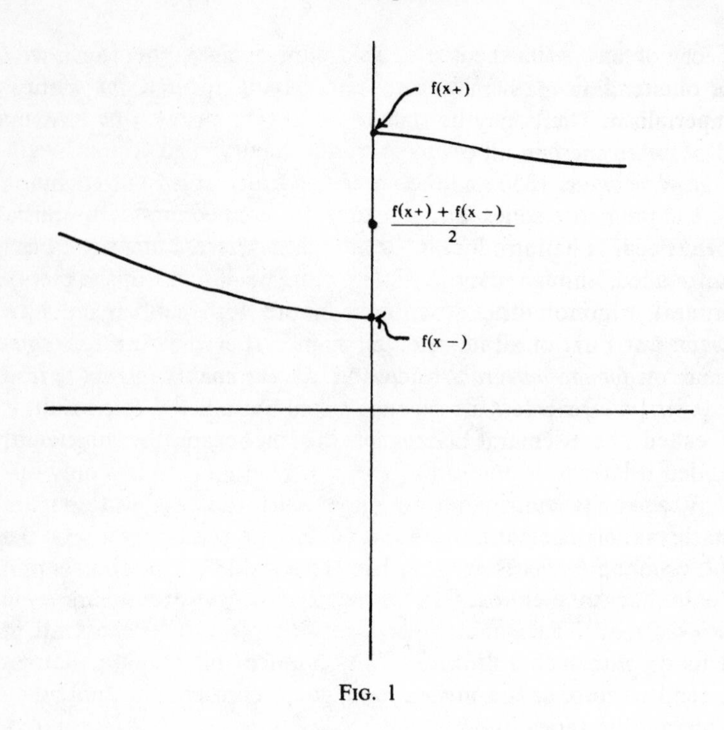

FIG. 1

By the way, this is the place where the notion of function as we know it in calculus texts was introduced. Prior to this, a function was something which could be defined by an analytic expression or by a geometrical curve. Let me mention parenthetically one story here. In the introduction to my book [2], I indicated that trigonometric series had considerable influence upon the development of analysis. There I mentioned the name of Dirichlet in connection with the notion of function; then the name of Riemann, who in his big paper, his main paper on trigonometric series, introduced his notion of integral; and also the name of Cantor, who was the first to study the so-called sets of uniqueness (I shall discuss them later) and whose attempt to characterize them, a problem which is still unsolved, led him to develop the theory of sets. After the book appeared, I had a conversation with an outstanding mathematician who accused me of mathematical imperialism. There may be something to that; there is an element of possessiveness in all of us.

Now between 1850 and 1860 appeared the memoir of Riemann on trigonometric series, and it was a very remarkable accomplishment because he introduced certain methods which have not been superseded, though they have been developed, in dealing with general trigonometric series. What are the main results of Riemann? First of all he proved the so-called Riemann-Lebesgue theorem: *for any integrable function* $f(x)$, *the coefficients* $c_n(f)$ *tend to* 0 (*as* $|n| \to \infty$). It is a very elementary but a very basic result. It is called the Riemann-Lebesgue theorem because Lebesgue extended it later to his notion of integral. That $c_n(f) \to 0$ is only one of Riemann's contributions. It is very curious that the main theorems of Riemann deal with general trigonometric series (which are not necessarily Fourier series). There is one result I would like to mention here. Suppose we have a trigonometric series $\Sigma c_n e^{inx}$. Riemann was the first to associate some kind of function with such a general series. Suppose, for example, that the c_n tend to zero, or are merely bounded. Riemann was the first to integrate the series formally. Consider the series, twice integrated, and call that sum $F(x)$:

$$c_0 \frac{x^2}{2} + \sum_{-\infty}^{\infty}{}' \frac{c_n}{(in)^2} e^{inx} = F(x)$$

where the central term, $n = 0$, is integrated separately. (The prime indicates the omission of that term.) He proved the following result: *if* $S_n(x_0)$ *tends to* c, *then if we form the function* $F(x)$ *and consider the second generalized derivative* (*nowadays called the Schwarz or Riemann-Schwarz derivative*)

$$\lim_{h \to 0} \frac{F(x+h) + F(x-h) - 2F(x)}{h^2},$$

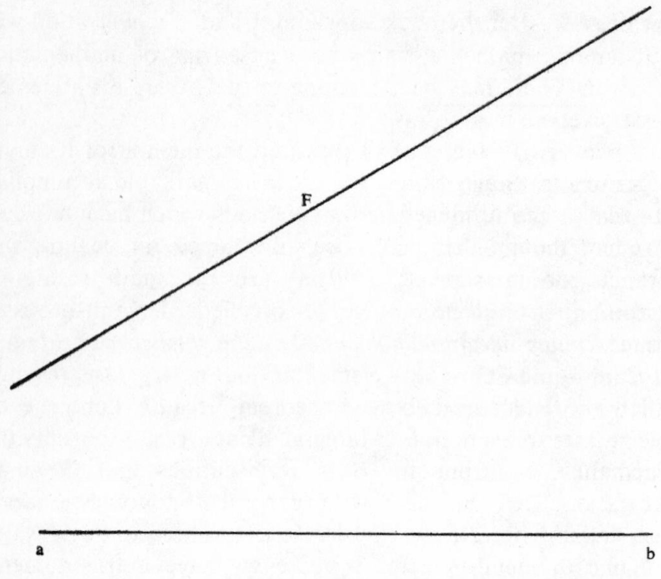

FIG. 2

that derivative exists at x_0 *and equals* c. This is called the first theorem of Riemann. What is the significance of this theorem? In fact, some people think that this was the beginning of the theory of distributions, because it associated with the most general trigonometric series a well-defined continuous function which is connected to the original series by differentiation. So, for example, if our initial series converges to zero in an interval (a, b), then the function F has a second Schwarz derivative equal to 0 in that interval, and therefore it is a familiar fact that F is linear on that interval [2, vol. 1, p. 23] (see Figure 2).

There is another theorem of Riemann here whose significance is less obvious, namely, if you merely assume that the coefficients c_n tend to 0, this implies that at every point x the function F satisfies

$$\lim_{h\to 0} \frac{F(x+h)+F(x-h)-2F(x)}{h} = 0. \tag{3}$$

What does (3) mean? It may be written

$$\frac{F(x+h)-F(x)}{h}-\frac{F(x)-F(x-h)}{h}\to 0.$$

In other words, under merely the condition $c_n \to 0$, the function F is *smooth*, in the following technical sense—the ratios from the right and from the left behave exactly in the same way, their difference tending to zero. Now these are the main contributions of Riemann, but increasingly there was quite a number of consequences which he himself did not develop. They came primarily from Cantor and Schwarz.

What I intend to mention in connection with Cantor is the following fact: Suppose that the original series converges to 0 in an interval (a, b), except possibly at one point c. What can be said about the series? From the preceding it follows that the function F is linear (see Figure 3) from a to c and linear from c to b, but it may have an angular point at $(c, F(c))$. However, the second theorem of Riemann shows that this is impossible. In other words,

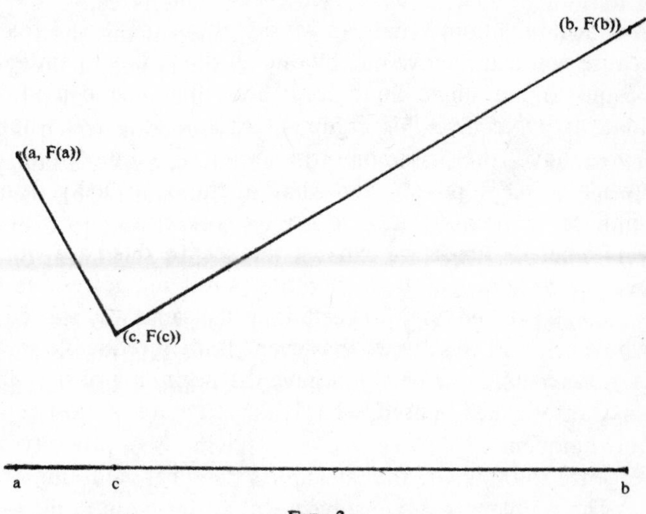

FIG. 3

the two segments $\overline{(a, F(a))(c, F(c))}$ and $\overline{(c, F(c))(b, F(b))}$ are continuations of each other; in other words, the function F is linear in the whole interval (a, b). Cantor was interested in the problem of uniqueness of trigonometric series. In other words, he looked at this problem: Suppose a trigonometric series $T = \Sigma c_n e^{inx}$ converges to 0 *everywhere*. The first theorem of Riemann easily shows that the coefficients must be all identically equal to 0. Why? Well, because the function F obtained by twice formally integrating is a linear function, and from this it readily follows that the initial series must be identically equal to 0 [2, vol. I, p. 326]. Now Cantor asked himself this question: What happens if the series converges to zero everywhere on $[0, 2\pi]$, except possibly at one point c? He immediately deduced by the bent-line argument mentioned above that the series T converges to 0 at the point c, and therefore everywhere. Of course if you have a finite number of points the argument is the same. So, in other words, every set consisting of a finite number of points is a *set of uniqueness*—if the series converges to 0 outside the set, it is automatically 0 identically.

Next step: Suppose that our exceptional set, about which we know nothing as regards the convergence of the series, has a single point of accumulation. What can we say about the series? It's still 0, because you can remove one by one all the isolated points until you come to the single limit point and then you can in turn eliminate it. Of course this argument can be repeated. Suppose that you have the trigonometric series T converging to 0 everywhere, except possibly at what might be called a finitely reducible set of points, that is to say by proceeding in this way a finite number of times, we exhaust the set. In this case we first remove the isolated points, then those points of the residue that have become isolated, and proceeding in the same way we remove the whole set, and this shows that every finitely reducible set is a set of uniqueness. Now here you have the beginning of the theory of sets. Cantor asked himself, what is the structure of most general sets of uniqueness? (These are sets having the property that convergence to 0 outside the set implies identical vanishing of the series.) The problem is still unsolved. But at this point in time—it's a very fashionable expression nowadays—the theory of sets truly arose and also the theory of transfinite numbers. Because clearly this process leads to the notion of transfinite numbers and they were introduced in this connection.

Here I'd like to mention the classical formula of Dirichlet for the partial sum of a Fourier series:

$$S_n(x) = \frac{1}{\pi} \int_{-\pi}^{\pi} f(x+t) D_n(t)\, dt \tag{4}$$

where $D_n(t)$ is the Dirichlet kernel:

$$D_n(t) = \frac{\sin(n+\frac{1}{2})t}{2 \sin \frac{1}{2} t}.$$

Observe that if we disregard the $\frac{1}{2}$ in $\sin(n+\frac{1}{2})t$ and associate the term $1/(2 \sin \frac{1}{2} t)$ with the function f, $S_n(x)$ is formally the Fourier (sine) coefficient of the function, namely $f(x+t)/(2 \sin \frac{1}{2} t)$. This is a very important fact, because from this it immediately follows by the theorem of Riemann about the coefficients tending to 0, that if a function f vanishes in an interval (a, b), then the n-th partial sum $S_n(x)$ tends to 0 in (a, b). In particular, if you have two functions coinciding in an interval, then their Fourier series behave much in the same way in the interval. They are *equiconvergent*—the difference of the partial sums of the two Fourier series converges to zero in that interval.

The problems of Fourier series lead also to considering the integral

$$-\frac{1}{\pi} \int_{-\pi}^{\pi} \frac{f(x+t)}{2 \tan \frac{1}{2} t}\, dt \tag{5}$$

whose similarity to the integral in (4) is clear. The value of the integral (5), defined as

$$-\frac{1}{\pi} \lim_{\epsilon \to 0} \left(\int_{-\pi}^{-\epsilon} + \int_{\epsilon}^{\pi} \right) \frac{f(x+t)}{2 \tan \frac{1}{2} t}\, dt,$$

is called the function *conjugate* to $f(x)$ and is usually denoted $\tilde{f}(x)$. It plays an important role in the theory of Fourier series and is represented by the so-called *conjugate series*—

$$\sum_{-\infty}^{+\infty} c_n(-i \operatorname{sgn} n)e^{inx}, \quad \text{where } \operatorname{sgn} n = \begin{cases} 1, & n > 0, \\ 0, & n = 0, \\ -1, & n < 0. \end{cases}$$

Basically, I'm practically through with the 19th century. There had been a number of other developments—only one of which I shall mention. It is the fact that the Fourier series of a continuous function can diverge. This was obtained by DuBois Reymond in 1876. He was the first to show that there exists a continuous function whose Fourier series may diverge at some point. In fact, it may even diverge on an infinite set of points. This discovery must have come as quite a shock to the analysts of that time, because at that time there was little notion of summability of Fourier series.

As we pass to the twentieth century, we come across two major developments. One of these is the creation of the Lebesgue theory of measure and integration. Lebesgue rebuilt the theory of Fourier series on his notion of integral and extended quite a number of known results to his more general situation. The other development is the concept of summability of Fourier series. Essentially it emerged primarily in a very special case, namely, in the work of Fejér, by considering the limits of the average of the partial sums,

$$\sigma_n = \frac{S_0 + S_1 + \cdots + S_n}{n+1}.$$

Of course representing $f(x)$ by its Poisson integral also means applying to the Fourier series of $f(x)$ a method of summability. Although this occurred even before the work of Fejér, the general notion of summability did not exist at that time. Only the function theoretical significance of the Poisson integral had been stressed. Since the σ_n have little connection with the theory of functions, Fejér's result brought the general notion of summability to the fore.

These two developments, namely, Lebesgue's measure and integration, and summability of series, changed the picture of Fourier series completely. I won't quote particular results, but you know that by using summability here we can restore order and a certain sense of aesthetics to the theory of Fourier series. This was done by quite a number of people. Let me mention only a few names. In the first place, there were de la Vallée-Poussin and Fatou, who essentially completed the rebuilding of the theory of Fourier series on the basis of Lebesgue integration. One should also mention here W. H. Young, who proved that every denumerable set, no matter what its structure, is automatically a set of uniqueness, and thereby disposed of transfinite induction. But basically the methods here are still continuations of the way it had been done by Lebesgue. This is more or less how the picture presented itself until, let's say, the 1920's as regards Fourier series.

I will start the more recent history of the twentieth century with a quick preview.

First of all, there was Hardy and Littlewood—contrary to spelling, in this context a single name. Hardy and Littlewood practically exhausted the topic. By using classical ideas, they essentially proved everything that could be proved without going beyond those ideas. (However, later they themselves went far beyond those ideas. Compare section 6 below.) Then there was Kolmogorov who showed that there exists an integrable function whose Fourier series diverges everywhere. This of course underlined the significance of summability in the theory of Fourier series. There were quite a number of also very outstanding mathematicians but I would like to mention two names of people who exerted considerable influence, though posthumously, and who died very early. One of them was Paley and the other was Marcinkiewicz. Paley, together with Hardy and Littlewood, extended the previous results by introducing certain new points of view partly based on the so-called complex methods. The significance of Marcinkiewcz cannot be possibly adequately described by the individual results. Although he made some basic results, solving problems difficult to solve, his main contribution was new methods. I think he had a very clear geometric intuition. I guess I have the feeling that the significance of his methods still has not been exhausted.

I will now discuss twentieth century developments in several areas.

2. CONJUGATE FUNCTIONS

That the conjugate function $\tilde{f}(x)$ exists almost everywhere for any f in L^2 is an early result of Lusin. The existence of $\tilde{f}$ for the most general integrable f is due to Privalov. In other words, the principal value integral defining $\tilde{f}$ has meaning for the most general functions integrable in the Lebesgue sense.

To Marcel Riesz in 1927 we owe the basic fact that *if $f(x)$ is in L^p, where p is finite and strictly greater than* 1, *then $\tilde{f}(x)$ is also in L^p and*

$$\|\tilde{f}\|_p \leqslant A_p \|f\|_p, \tag{6}$$

where A_p depends on p only. (The result had of course been known for $p = 2$. This follows immediately from Parseval's theorem which is mentioned in section 5 below.) From (6) it easily follows that the partial sums $S_n[f]$ tend to f in the metric L^p,

$$\|S_n - f\|_p \to 0 \text{ as } n \to \infty. \tag{7}$$

(We also have the parallel result that

$$\|\tilde{S}_n - \tilde{f}\|_p \to 0,$$

where $\tilde{S}_n$ denotes the partial sum of the conjugate series.)

Inequality (7) fails for $p = 1$ or $p = \infty$. The case $p = 1$ deserves special mention since it leads to the very important notion of weak integrability. Kolmogorov (1922) has shown that although $\tilde{f}(x)$ need not be integrable it has a weaker property—the set of points x where $|\tilde{f}(x)| > y$ has measure less than a constant multiple of $\|f\|_1/y$.

3. COMPLEX METHODS

Again in this connection I should mention the collective name: Russian school. Collective here is quite accidental. What is the Russian school? The Russian school was essentially created by Lusin. He had very outstanding students. One of them was Kolmogorov whom I have just mentioned. Then there were Menšhov, Privalov, and a few others. What is the significance of the Russian school? What was their approach? They developed the so-called complex method—namely associating a trigonometric series with a power series. For example, say we have a general trigonometric series $\Sigma c_n e^{inx}$. Suppose that $c_{-n} = \bar{c}_n$; then this can be written also in the form $\frac{1}{2}a_0 + \Sigma_{n=1}^{\infty}(a_n \cos nx + b_n \sin nx)$ and so represents the real part of the power series $\frac{1}{2}a_0 + \Sigma_{n=1}^{\infty}(a_n - ib_n)e^{inx}$. In other words, treating trigonometric series as real parts of power series immediately makes applicable methods of complex variables. You know that analytic functions have quite a number of properties and one of the main achievements of the Russian school was a very extensive study of boundary value properties of analytic functions defined in the unit disc. What are the main results? One of them was the result of Privalov concerning $\tilde{f}$ I have just mentioned in connection with the conjugate functions above. There was another development introduced by Lusin himself. Of course, prior to him, people normally considered either a radial or a non-tangential approach to the boundary. Now suppose you have, let's say, $\varphi(z)$, defined inside the unit circle. Lusin was the first to introduce a different notion, namely, the so-called *area function*. Fix δ, $0 < \delta < 1$ (see Figure 4), and let A_θ be the shaded region, then the area function is given by

$$S(\theta) = \int_{A_\theta} |\varphi'(z)|^2 \, d\sigma.$$

He showed the finiteness of this function for almost all θ's if $\varphi(z) = \Sigma c_n z^n$ and $\Sigma |c_n|^2 < \infty$. By now we know that for the most general trigonometric series the finiteness of this integral is

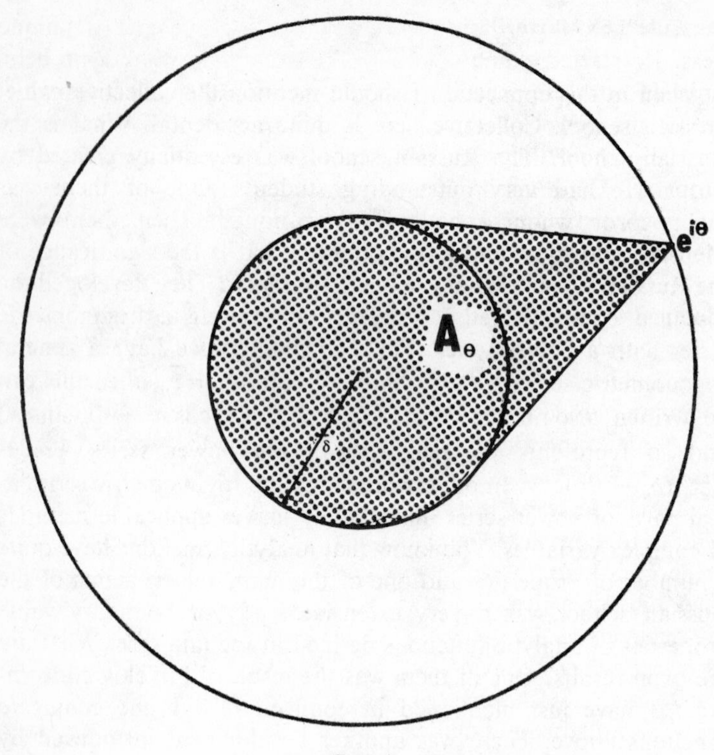

FIG. 4

equivalent to the existence of the non-tangential limit (up to a set of θ's of measure 0); and the area function has become a very important tool for the study of trigonometric series.

4. SETS OF UNIQUENESS AND MULTIPLICITY

Here let me mention what were the contributions of Menshov. They dealt with the problems of uniqueness. W. H. Young, extending previous results, proved that every denumerable set is a set of uniqueness. A very natural question was, how about sets which were not? Menshov was the first to prove that there exist sets of measure zero—even perfect sets—which are not sets of uniqueness. He started a line of problems which is very far from being finished now. He proved that there are sets of uniqueness which are of the power of the continuum (but necessarily of measure zero). But we still don't know what sets of measure zero have that property. He himself showed that there exist certain sets of measure zero which are not sets of uniqueness. In other words, they are *sets of multiplicity*, that is to say, there do exist trigonometric series converging to zero outside the set without being identically zero.

By the way, this is to underline a development which has had comparatively less success than other developments. Dirichlet and Riemann pushed trigonometric series in the direction of the theory of a function of a real variable. The work of the Russian school, which by the way was based upon the work of Fatou, pushed trigonometric series in the direction of complex variables. However, trigonometric series have yet more connections with other fields—particularly with the theory of numbers. If we look at the trigonometric series $\Sigma c_n e^{inx}$, we observe that what matters here is the behavior of the product nx taken modulo 2π. In other words, as regards their behavior, trigonometric series have a very strong connection with Diophantine properties of numbers. So far, there are very brilliant individual results,† but this direction has been much less successful (I guess that's my own personal feeling) than the other developments and probably the future belongs to it. We have quite a number of applications of trigonometric series to the theory of numbers; for example one can see this in the work of

†A good example is the following theorem of Salem and Zygmund: *Given a real* $\xi > 1$ *form a "Cantor set"* C_ξ *on* $[0, 2\pi]$ *by removing an open interval of length* $2\pi/\xi$ *from the middle, then removing open intervals of length* $2\pi/\xi^2$ *from the middle of each of the two remaining closed intervals, and iterating this process countably often. Then* C_ξ *(which has measure* 0) *is a set of uniqueness if and only if* ξ *is an algebraic integer all of whose conjugates have modulus less than* 1 (i.e., $\xi^n + a_1\xi^{n-1} + \cdots + a_n = 0$ for some integers $a_1, a_2, \ldots, a_n$—n is minimal, and the other $n-1$ roots of $x^n + \cdots + a_n = 0$ have absolute value less than one). In particular, the classical Cantor set C_3 is a set of uniqueness, while the "intuitively thinner" $C_{5/2}$ is not. I find this a most surprising result.—*Editor*.

Vinogradov. But there are comparatively few applications of the theory of numbers to trigonometric series.

5. INTERPOLATION OF OPERATORS

The theorem of Parseval says,

$$\sum_{n=-\infty}^{\infty} |c_n(f)|^2 = \frac{1}{2\pi}\int_0^{2\pi} |f|^2\, dx.$$

We have also the obvious inequality

$$|c_n(f)| = \left|\frac{1}{2\pi}\int_0^{2\pi} f(t)e^{-int}\, dt\right| \leqslant \frac{1}{2\pi}\int_0^{2\pi} |f|\, dx.$$

So we know something about the Fourier coefficients of functions in L^2 and also functions in L^1. The problem is: what can you say about the classes L^p, where p is not necessarily 1 or 2? The basic result here is the so-called Hausdorff–Young theorem. Using these notions of norms:

$$\|c\|_p = \left(\sum_{n=-\infty}^{\infty} |c_n|^p\right)^{1/p}, \qquad \|f\|_q = \left(\frac{1}{2\pi}\int_0^{2\pi} |f|^q\, dx\right)^{1/q},$$

it basically says that

$$\|c\|_{p'} \leqslant \|f\|_p,$$

where p lies between 1 and 2 and $p' = p/(p-1)$ (so that $1/p + 1/p' = 1$) is the index conjugate to p. This is one part of the Hausdorff–Young theorem. I remember when I studied the paper of Hausdorff (I never read Young—Young proved it only for special values of p), I was always very much impressed by it, but I never understood it. It was exceedingly technical. One could admire the skill and depth of the argument, but one couldn't really (this is my own feeling) understand the interpolation to general p.

A new proof of this theorem was due to Marcel Riesz. Again this was a very important development of the theory of functional analysis. Riesz points out that the Hausdorff-Young theorem basically is merely a special case of a very general theorem concerning interpolation of linear operators. M. Riesz's proof was not only important because of this particular result, but because it gave a completely general point of view upon the applications of the theory of operators. Now, if we get a couple of results about some inequalities of norms (e.g., $\|c\|_2 \leqslant \|f\|_2$, $\|c\|_\infty \leqslant \|f\|_1$), we almost automatically interpolate (getting $\|c\|_{p'} \leqslant \|f\|_p$), following M. Riesz. To my mind (one can have of course personal feelings), this is probably one of the most interesting, the most outstanding results in analysis—if only because it shows a very general point of view.

Here comes a result in this connection under the name of Marcinkiewicz. Important as it is, M. Riesz's theorem is not always applicable because sometimes the result of taking a formal interpolation between two norm inequalities is correct without those norm inequalities being valid. In other words, the function need not satisfy the hypotheses of M. Riesz and yet his conclusion may still be valid. Marcinkiewicz obtained a much more general theorem. It's interesting not only because of the generality, but also because essentially in the case of the proof of Marcel Riesz complex variables play too important a role. Really you don't know what's happening. You apply the three circle theorem and the result comes out neatly. But what actually is going on? You don't know. Now, the proof of Marcinkiewicz has the advantage that actually you see with naked eyes what various parts of functions contribute to the value of the operation.

6. THE MAXIMAL THEOREM OF HARDY AND LITTLEWOOD

This is a result of considerable and lasting importance, introducing a point of view which permeates many developments in analysis. Consider a locally integral function $f(x)$ of a single variable and its mean value

$$\frac{1}{2h}\int_{x-h}^{x+h} f(t)\,dt.$$

We know that, as h tends to 0, this tends to $f(x)$ (almost everywhere for f merely in L^1); but Hardy and Littlewood had the idea of considering not the limit, but

$$\sup_{h>0}\frac{1}{2h}\int_{x-h}^{x+h} |f(t)|\,dt.$$

This is the so-called maximal function of Hardy and Littlewood. They obtained very remarkable results here—as the French say, "chapeau bas." This function is much easier to apply than is the limit. They proved a number of basic properties of this maximal function—properties which later on have been extended to higher dimensions and we know play a very important role in modern analysis.

By the way, I should mention one fact which probably has been overlooked. Essentially the idea of considering the sup instead of the limit—in other words, considering let's say the expression $S_{n(x)}(x)$ instead of $\lim_{n\to\infty} S_n(x)$, appears in an old paper of Hermann Weyl where he considers convergence of orthogonal series [1]. Unfortunately that paper doesn't sufficiently exploit the brilliance of this idea.

I mentioned a number of theorems of Hardy and Littlewood without stating them. Let me not go into details here, but I would like at the end to mention one development in which I became very much interested when I was young. And this unfortunately didn't lead to anything. It is the problem of convergence and behavior of general orthogonal series. Between 1910 and 1930, quite a number of papers had been written about general orthogonal series $\Sigma_{n=1}^{\infty} c_n\phi_n(x)$, usually under the hypothesis that $\Sigma_{n=1}^{\infty}|c_n|^2 < \infty$. What are the properties of such series? If my recollection is correct, the expectation was that by solving the problem of convergence of general orthogonal series, we would automatically solve the corresponding problem for trigonometric series, which of course was the central one—the one everybody wanted to solve. Somehow at that time it was not properly understood that we couldn't expect very much in that direction, because the notion of general orthogonal sequence is independent of the ordering. In other words, if we proved something about general orthogonal series, we automatically could prove corresponding results for trigonometric series with arbitrary ordering. In other words, we would have proved something about the unconditional convergence of Fourier series. The topic has appeared explicitly only recently. That is one of the topics which probably will be studied! What are the properties, let's say, of convergence of a trigonometric series if we change the order in an arbitrary way? There are some negative results. I think it was proved by Zahorski in 1960 that in the case of L^2 already the Fourier series may diverge almost everywhere for the proper rearrangement of terms.

But definitive results are very few. Now we may ask; what is the purpose of all those investigations about unconditional convergence? It's a matter of taste. So what? I like it and that's the end.

REFERENCES

1. Jerosch, F., and H. Weyl, "Uber die Konvergenz von Reihen, die nach periodischen Funktionen fortschreiten", *Math. Ann.*, **66** (1909), 67–80.
2. Zygmund, A., *Trigonometric Series*, 2nd ed., Cambridge University Press, New York, 1959.

DEVELOPMENTS RELATED TO THE A.E. CONVERGENCE OF FOURIER SERIES

Richard A. Hunt

0. INTRODUCTION

The most striking result in the field is obviously L. Carleson's proof of Lusin's conjecture. In 1913 Lusin [11] conjectured that the Fourier series of every L^2 function converges a.e. and Carleson [3] provided the proof in 1966. The proof is as complicated as the statement is simple and we will not try to explain the proof here. Instead, we will devote our attention to the development of results on which it is based.

It is interesting that Carleson's ingenious construction was achieved by using tools which are standard in classical harmonic analysis. That is, it is based on Parseval's formula and on L^p and weak L^p estimates of the maximal conjugate function, the Hardy-Littlewood maximal function, and a maximal function which is associated with the partial sums of the Fourier series of f.

It is not surprising that the conjugate function plays an important role in Carleson's proof. Lusin based his conjecture on his result that the conjugate function of f exists and is in L^2 for every f

in L^2. The work of Calderón and Zygmund [2] provides an important connection between the conjugate function and the Hardy-Littlewood maximal function. Their results and a variant of part of their proof are used in Carleson's proof.

L^p estimates of the maximal conjugate function and the Hardy-Littlewood maximal function are obtained from weak L^p estimates and the Marcinkiewicz Interpolation Theorem. The actual convergence of the partial sums of the Fourier series is a consequence of an important general principle. That is, under very general conditions on the L^p operators T_n, the sequence $T_n f$ converges a.e. for all $f \in L^p$ if and only if the associated maximal operator $T^*f = \sup|T_n f|$ satisfies a weak L^p inequality.

In this paper we will illustrate the development by presenting proofs as they stand today. (For an historical development and a brief indication of how the results fit into Carleson's proof, see Hunt [8].) For the sake of simplicity we will prove dyadic analogues of some of the results. That is, the dyadic Hardy-Littlewood maximal function will be used in place of the ordinary one and Walsh-Fourier series will be used in place of trigonometric Fourier series. This allows us to see the essential ideas of the development without many of the purely technical details.

1. WEAK L^P INEQUALITIES AND THE MARCINKIEWICZ INTERPOLATION THEOREM

L^p and weak L^p estimates both depend on the distribution function. Recall that the distribution function of f is defined to be

$$\lambda_f(y) = m\{x : |f(x)| > y\}, \quad y > 0. \tag{1.1}$$

All functions will be assumed to be Lebesgue measurable on [0, 1] and m will denote Lebesgue measure on [0, 1].

The L^p norm of f obviously depends in some way on λ_f. The following result is more specific and its proof illustrates how estimates of λ_f can be used to estimate $\|f\|_p$:

$$p\int_0^\infty y^{p-1}\lambda_f(y)\,dy = \int_0^1 |f(x)|^p dx, \quad p \geqslant 1. \tag{1.2}$$

Proof:

$$p\int_0^\infty y^{p-1}\lambda_f(y)\,dy = p\int_0^\infty y^{p-1}\left[\int_0^1 \chi_{\{|f(x)|>y\}}(x,y)\,dx\right]dy,$$

where χ_E denotes the characteristic function of the set E. Application of the Fubini–Tonelli Theorem to the above double integral yields

$$\int_0^1\left[p\int_0^\infty y^{p-1}\chi_{\{|f(x)|>y\}}(x,y)\,dy\right]dx$$

$$= \int_0^1\left[p\int_0^{|f(x)|} y^{p-1}\,dy\right]dx$$

$$= \int_0^1 |f(x)|^p\,dx,$$

and (1.2) follows.

The operators that we will consider are all either linear or a pointwise supremum of linear operators. All such operators are sublinear. That is, $T(f+g)$ and $T(cf)$ are defined (a.e.) whenever Tf and Tg are and

$$|T(f+g)| \leqslant |Tf| + |Tg| \text{ and } |T(cf)| = |c|\cdot|Tf|. \tag{1.3}$$

If T is sublinear, note that

$$\{|Tf| \leqslant y\} \cap \{|Tg| \leqslant y\} \subset \{|T(f+g)| \leqslant 2y\}.$$

It follows that

$$\lambda_{T(f+g)}(2y) \leqslant \lambda_{Tf}(y) + \lambda_{Tg}(y). \tag{1.4}$$

This version of the triangle inequality is quite useful.

If T is a continuous sublinear operator on L^p $(1 \leqslant p \leqslant \infty)$, we have

$$\|Tf\|_p \leqslant C\|f\|_p (C \text{ independent of } f), \tag{1.5}$$

and we say T is of type (p, p). If $1 \leqslant p < \infty$ and

$$y^p \lambda_{Tf}(y) \leqslant C^p \|f\|_p^p \text{ for all } y > 0 \text{ and } f \in L^p, \tag{1.6}$$

we say T is of weak type (p, p).

Note that

$$y^p \lambda_{Tf}(y) = \int_{\{|Tf| > y\}} y^p dx$$
$$\leqslant \int_0^1 |Tf(x)|^p dx.$$

Hence, if T is of type (p, p) with constant C, then T is of weak type (p, p) with constant C. The converse is not true, but the following interpolation theorem holds:

THEOREM: *Suppose T is sublinear and*

$$y^{p_j} \lambda_{Tf}(y) \leqslant C_j^{p_j} \int_0^1 |f(x)|^{p_j} dx$$

for all $f \in L^{p_j}$ and all $y > 0$, $j = 0, 1$, and $1 \leqslant p_0 < p_1 < \infty$. In case $p_1 = \infty$ we replace the corresponding weak type hypothesis by the condition $\|Tf\|_\infty < C_1 \|f\|_\infty$. Then

$$\int_0^1 |Tf(x)|^p dx \leqslant p 2^p C_0^{p_0((p_1 - p)/(p_1 - p_0))} C_1^{p_1((p - p_0)/(p_1 - p_0))}$$
$$\times \left[\frac{1}{p - p_0} + \frac{1}{p_1 - p} \right] \int_0^1 |f(x)|^p dx$$

for all $f \in L^p$, $p_0 < p < p_1$.

Proof: For fixed $y > 0$, set

$$f^y(x) = \begin{cases} f(x), \text{ if } |f(x)| \leqslant Ay, \\ 0, \text{ otherwise,} \end{cases}$$

and

$$f_y(x) = \begin{cases} 0, \text{ if } |f(x)| \leqslant Ay, \\ f(x), \text{ otherwise,} \end{cases}$$

where $A = C_0^{p_0/(p_1-p_0)} C_1^{-p_1/(p_1-p_0)}$.
Then $f(x) = f_y(x) + f^y(x)$. Since T is sublinear we have

$$\lambda_{Tf}(2y) \leqslant \lambda_{Tf_y}(y) + \lambda_{Tf^y}(y).$$

In case $p_1 < \infty$, the hypothesis yields

$$\lambda_{Tf}(2y) \leqslant C_0^{p_0} y^{-p_0} \int_0^1 |f_y(x)|^{p_0}\, dx + C_1^{p_1} y^{-p_1} \int_0^1 |f^y(x)|^{p_1}\, dx$$

$$= C_0^{p_0} y^{-p_0} \int_0^1 \chi_{\{|f(x)| > Ay\}}(x, y) |f(x)|^{p_0}\, dx$$

$$+ C_1^{p_1} y^{-p_1} \int_0^1 \chi_{\{|f(x)| \leqslant Ay\}}(x, y) |f(x)|^{p_1}\, dx.$$

To complete the proof, multiply this above inequality by $p2^p y^{p-1}$, integrate with respect to y from 0 to ∞, and use the Fubini-Tonelli theorem as before.

In case $p_1 = \infty$ we have $\|f^y\|_\infty \leqslant Ay = C_1^{-1} y$ and $\|Tf^y\|_\infty \leqslant C_1 \|f^y\|_\infty \leqslant y$. This implies that $\lambda_{Tf^y}(y) = 0$, so the p_1 part of the above argument drops out.

(The above is a special case of the Marcinkiewicz Interpolation Theorem. See [15; Vol. II, p. 112] for a more general result.)

2. THE HARDY-LITTLEWOOD MAXIMAL FUNCTION

Intervals of the form $(k2^{-n}, (k+1)2^{-n})$, $n \geqslant 0$, $k = 0, \ldots, 2^n - 1$, are called dyadic subintervals of $[0, 1]$. ω will always denote a dyadic subinterval of $[0, 1]$ and $|\omega|$ will denote its length. An important property of dyadic intervals is that $\omega_1 \cap \omega_2 \neq \phi$ implies either $\omega_1 \subset \omega_2$ or $\omega_2 \subset \omega_1$.

The (dyadic) Hardy-Littlewood maximal function is defined by

$$\Lambda f(x) = \sup\left\{ \frac{1}{|\omega|} \int_\omega |f(t)|\,dt : x \in \omega \right\}. \tag{2.1}$$

(Ordinarily, Λf is defined by using intervals I with center x in place of the dyadic intervals ω.)

Let us consider the set $E_y = \{x : \Lambda f(x) > y\}$, $y > 0$. In case $\int_0^1 |f(t)|\,dt > y$ we have $E_y = [0, 1]$. In case $\int_0^1 |f(t)|\,dt \leqslant y$ it is possible to find a sequence $\{\omega_j\}_{j \geqslant 1}$ such that

$$\begin{cases} \omega_j \cap \omega_k = \phi \quad \text{if} \quad j \neq k, \\ E_y = \bigcup_{j \geqslant 1} \omega_j, \quad \text{and} \\ y < \dfrac{1}{|\omega_j|} \displaystyle\int_{\omega_j} |f(t)|\,dt \leqslant 2y, \quad j \geqslant 1. \end{cases} \tag{2.2}$$

This is easy to see in the dyadic case. That is, for each $x \in E_y$ let ω_x denote the largest ω with $x \in \omega$ and $(1/|\omega|)\int_\omega |f(t)|\,dt > y$. Since $\omega_x \subsetneq [0, 1]$, there is a unique dyadic interval $\omega_x^* \subset [0, 1]$ such that $\omega_x \subset \omega_x^*$ and $|\omega_x^*| = 2|\omega_x|$. Then $(1/|\omega_x^*|)\int_{\omega_x^*} |f(t)|\,dt \leqslant y$, so

$$\begin{aligned} \frac{1}{|\omega_x|} \int_{\omega_x} |f(t)|\,dt &= \frac{2}{|\omega_x^*|} \int_{\omega_x} |f(t)|\,dt \\ &\leqslant \frac{2}{|\omega_x^*|} \int_{\omega_x^*} |f(t)|\,dt \leqslant 2y. \end{aligned}$$

Clearly, $E_y = \cup_{x \in E_y} \omega_x$. Since $\omega_x \cap \omega_{\bar{x}} \neq \phi$, $x, \bar{x} \in E_y$, implies $\omega_x \equiv \omega_{\bar{x}}$, we can obtain the desired sequence $\{\omega_j\}_{j \geqslant 1}$ by discarding duplicates from the collection $\{\omega_x\}_{x \in E_y}$.

Using (2.2), we have

$$mE_y = \sum_{j \geq 1} |\omega_j| \leq \sum_{j \geq 1} \frac{1}{y} \int_{\omega_j} |f(t)|\,dt$$

$$= \frac{1}{y} \int_{E_y} |f(t)|\,dt \leq \frac{1}{y} \|f\|_1. \tag{2.3}$$

Also, $mE_y \leq y^{-1}\|f\|_1$ if $\int_0^1 |f(t)|\,dt > y$, so Λ is of weak type (1, 1). Clearly, $\|\Lambda f\|_\infty \leq \|f\|_\infty$, so the Marcinkiewicz Interpolation Theorem yields the following result:

THEOREM:

$$\|\Lambda f\|_p \leq \left[p2^p/(p-1) \right]^{1/p} \|f\|_p, \quad 1 < p < \infty.$$

In the case of the ordinary Hardy-Littlewood maximal function there are several variations of (2.2) which hold and give the weak type (1, 1) result of (2.3). For example, see Besicovitch [1] or Stein [13]. The original proof of the L^p result given by Hardy and Littlewood [7] is quite different from the proof we have given. By the way, don't miss their introduction of the problem in terms of cricket averages!

The results we have proved for the Hardy-Littlewood maximal function provide a nice opportunity to illustrate the connection between the a.e. convergence of a sequence of L^p operators and a weak type (p, p) inequality for the corresponding maximal function. To do this, let $T_n f(x) = \frac{1}{|\omega|} \int_\omega f(t)\,dt$, where $x \in \omega$ and $|\omega| = 2^{-n}$, $n = 0, 1, \ldots$. The associated maximal function is $\sup_n |T_n f(x)| \leq \Lambda f(x)$.

If g is continuous on [0, 1], it is clear that $\lim_{n \to \infty} T_n g(x) = g(x)$, uniformly in x. That is, the sequence $T_n g$ converges a.e. for all g in a dense subset of L^1. This fact may be combined with the weak type (1, 1) estimate of the maximal function to obtain the following, which is a special case of Lebesgue's theorem:

THEOREM: *If $f \in L^1$, then $\lim_{n\to\infty} T_n f = f$ a.e.*

Proof: For fixed $f \in L^1$ it is enough to show that given any $\epsilon > 0$ we have $m\{\limsup_{n\to\infty}|T_n f - f| > \epsilon\} < \epsilon$. To do this choose a continuous function g such that $\|f - g\|_1 < \epsilon^2/6$. Since $\limsup_{n\to\infty}|T_n g - g| = 0$ (a.e.), we have

$$\begin{aligned} m\Big\{\limsup_{n\to\infty}|T_n f - f| > \epsilon\Big\} &\leqslant m\Big\{\limsup_{n\to\infty}|T_n f - T_n g| > \epsilon/3\Big\} \\ &\quad + m\Big\{\limsup_{n\to\infty}|T_n g - g| > \epsilon/3\Big\} \\ &\quad + m\{|g - f| > \epsilon/3\} \\ &\leqslant m\{\Lambda(f - g) > \epsilon/3\} + (\epsilon/3)^{-1}\|f - g\|_1 \\ &\leqslant 2(\epsilon/3)^{-1}\|f - g\|_1 < \epsilon. \end{aligned}$$

The above proof may be applied to Fourier series. For example, let $S_n f$ denote the nth partial sum of the Fourier series of f. If g is a trigonometric polynomial, it is clear that $S_n g \to g$ a.e. Since the trigonometric polynomials are dense in L^p ($p < \infty$), we would have that $S_n f \to f$ a.e. for all $f \in L^p$ if

$$m\Big\{\sup_n |S_n f| > y\Big\} \leqslant C_p^p y^{-p}\|f\|_p^p, \tag{2.4}$$

C_p independent of $f \in L^p$ and $y > 0$.

Carleson proved (2.4) in the case $p = 2$. Actually, we have the stronger result that

$$\|\sup_n |S_n f|\,\|_p \leqslant C_p\|f\|_p, \quad 1 < p < \infty. \tag{2.5}$$

Hunt [9] proved (2.5) by combining a slight modification of Carleson's proof with a generalization of the Marcinkiewicz Interpolation Theorem.

It is interesting that (2.4) with $p = 2$ is a necessary condition for the a.e. convergence of $S_n f$ to f for every $f \in L^2$. This is a result of Calderón. (See [15; Vol. II, p. 165].) In fact, under very general conditions on the L^p operators T_n, if the maximal function $T^*f = \sup_n |T_n f|$ is finite a.e. for all $f \in L^p$, then T^* must be of weak type (p, p). (See Stein [14] and the interesting discussion of the general principle given in Garsia [6].) Thus, in many important cases a weak type (p, p) inequality for the maximal operator T^* is necessary and sufficient for the a.e. convergence of $T_n f$ for every $f \in L^p$.

3. THE CALDERÓN-ZYGMUND DECOMPOSITION AND THE CONJUGATE FUNCTION

The Dirichlet formula for the nth partial sum of the Fourier series of f is

$$S_n f(x) = \frac{1}{\pi} \int_{-\pi}^{\pi} f(t) \frac{\sin(n + \frac{1}{2})(t - x)}{2 \sin \frac{1}{2}(t - x)} \, dt$$

$$= \frac{1}{\pi} \int_{-\pi}^{\pi} f(t) \left[\frac{\sin n(t - x)}{2 \tan \frac{1}{2}(t - x)} + \tfrac{1}{2} \cos n(t - x) \right] dt. \quad (3.1)$$

This is closely related to the conjugate function of f,

$$\tilde{f}(x) = \lim_{\epsilon \to 0^+} \tilde{f}_\epsilon(x)$$

$$= \lim_{\epsilon \to 0^+} \frac{-1}{\pi} \int_{\epsilon \leq |t-x| \leq \pi} f(t) \frac{dt}{2 \tan \frac{1}{2}(t - x)} . \quad (3.2)$$

In fact,

$$S_n f(x) = -\cos nx (f(\cdot) \sin n(\cdot))^{\sim}(x)$$

$$+ \sin nx (f(\cdot) \cos n(\cdot))^{\sim}(x)$$

$$+ \frac{1}{2\pi} \int_{-\pi}^{\pi} f(t) \cos n(t - x) dt. \quad (3.3)$$

In order to illustrate the techniques involved in the analysis of $S_n f$ and $\tilde{f}$ it is convenient to consider their dyadic analogues. That is, we consider Walsh-Fourier series in place of trigonometric Fourier series (See Fine [5] or the appendix to this paper.)

We identify $[0, 1]$ with the Walsh group 2^ω. That is, $x = \sum_{j=0}^{\infty} \eta_j 2^{-j-1}$, $\eta_j = 0$ or 1, corresponds to $(\eta_0, \eta_1, \ldots) \in 2^\omega$. If $t = \sum_{j=0}^{\infty} \tau_j 2^{-j-1}$, we define $x \dot{+} t = \sum_{j=0}^{\infty} |\eta_j - \tau_j| 2^{-j-1}$. Then $\{x \dot{+} t : t \in \omega\}$ is a dyadic interval of the same length as ω. If $S_n f(x)$ denotes the nth partial sum of the Walsh Fourier series of f we have that $S_{2^n} f(x) = (1/|\omega|) \int_\omega f(t)dt$, where $x \in \omega$ and $|\omega| = 2^{-n}$. Our previous result showed that $S_{2^n} f \to f$ a.e. for $f \in L^1$, so the Walsh functions $\{w_n\}_{n \geqslant 0}$ form a complete O.N. set. Hence, we have Parseval's equation

$$\sum_{n \geqslant 0} |c_n|^2 = \int_0^1 |f(t)|^2 \, dt. \tag{3.4}$$

Corresponding to (3.3), we define the operator S_n^* by

$$S_n^*(f)(x) = w_n(x) S_n(w_n f)(x). \tag{3.5}$$

Thus, S_n^* is an analogue of the conjugate operator.

We have $S_n^* f(x) = \int_0^1 f(t) D_n^*(x \dot{+} t)dt$, where $D_n^*(t) = w_n(t) D_n(t)$ and D_n is the nth Dirichlet kernel. It can be shown that $D_n^*(x \dot{+} t)$ is constant as t varies over any ω which does not contain x. This corresponds to the fact that $1/2 \tan \frac{1}{2}(x - t)$ is nearly constant as t varies over any interval I with $|I| \leqslant \operatorname{dist}(x, I)$.

Since $|w_n(x)| = 1$, we see from (3.5) that S_n^* and S_n are simultaneously of type (p, p) or of weak type (p, p). In particular,

$$\|S_n^* f\|_2 = \|S_n(w_n f)\|_2 \leqslant \|w_n f\|_2 = \|f\|_2. \tag{3.6}$$

Let us use the Calderón-Zygmund decomposition to show that S_n^* is of weak type $(1, 1)$. Recalling (2.2), the characterization of

$\{\Lambda f > y\} = \cup_{j \geq 1}\omega_j$, we define

$$g(x) = \begin{cases} \dfrac{1}{|\omega_j|}\displaystyle\int_{\omega_j} f(t)dt, & \text{if } x \in \omega_j, \\ f(x), & \text{if } x \not\in \displaystyle\bigcup_{j \geq 1} \omega_j, \end{cases}$$

and $b(x) = f(x) - g(x)$.

If $x \in \cup\, \omega_j$, then (2.2) implies $|g(x)| \leq 2y$. If $x \not\in \cup\, \omega_j$, then $x \in \omega$ implies $\left|(1/|\omega|)\int_\omega f(t)dt\right| \leq y$. Since these averages converge to f a.e., we have that $|f(x)| \leq y$ for a.e. $x \not\in \cup_{j \geq 1}\omega_j$. Hence, $\|g\|_\infty \leq 2y$. It is also clear that $\int |g| \leq \int |f|$. Hence, using (3.6) we have

$$m\{|S_n^*g(x)| > y/2\} \leq \frac{\int |S_n^* g|^2}{(y/2)^2}$$

$$\leq \frac{4\int |g|^2}{y^2} \leq \frac{8\int |g|}{y} \leq \frac{8\int |f|}{y}. \tag{3.7}$$

Note that $b(t)$ is zero if $x \not\in \cup_{j \geq 1}\omega_j$ and $\int_{\omega_j} b(t)dt = 0$, $j \geq 1$. Since $D_n^*(x \dotplus t)$ is constant as t ranges over ω_j if $x \not\in \cup\, \omega_j$, we have

$$S_n^* b(x) = \sum_{j \geq 1} \int_{\omega_j} b(t) D_n^*(x \dotplus t)dt$$

$$= \sum_{j \geq 1} \left[\frac{1}{|\omega_j|}\int_{\omega_j} D_n^*(x \dotplus t)dt\right]\left[\int_{\omega_j} b(t)dt\right] = 0,$$

if $x \notin \cup \omega_j$. Therefore, recalling (2.3), we have

$$m\{|S_n^*b| > y/2\} \leqslant m\{|S_n^*b| > 0\}$$

$$\leqslant m(\cup \omega_j) \leqslant \frac{1}{y}\int_0^1 |f(t)|dt. \tag{3.8}$$

Using (1.4) we see that (3.7) and (3.8) imply

$$m\{|S_n^*f| > y\} \leqslant 9y^{-1}\|f\|_1, \quad y > 0. \tag{3.9}$$

From (3.6), (3.9), and the Marcinkiewicz Interpolation Theorem we obtain

$$\|S_n^*f\|_p \leqslant C_p\|f\|_p, \quad 1 < p \leqslant 2. \tag{3.10}$$

We use the following duality argument to extend this inequality to $p > 2$: If $f \in L^p$ and $g \in L^q$, $(1/p) + (1/q) = 1$, $2 < p < \infty$, then $1 < q < 2$, so

$$\left|\int_0^1 S_n^*f(x)g(x)dx\right|$$

$$= \left|\int_0^1 \left[\int_0^1 f(t)D_n^*(x \dotplus t)dt\right] g(x)dx\right|$$

$$= \left|\int_0^1 f(t)\left[\int_0^1 g(x)D_n^*(x \dotplus t)dx\right]dt\right|$$

$$= \left|\int_0^1 f(t)S_n^*g(t)dt\right|$$

$$\leqslant \|f\|_p\|S_n^*g\|_q \leqslant C_q\|f\|_p\|g\|_q.$$

Since $\|S_n^*f\|_p = \sup\left\{\left|\int_0^1 S_n^*f \cdot g\right| : \|g\|_q \leqslant 1\right\}$, it follows that

$\|S_n^*f\|_p \leqslant C_q\|f\|_p$. Hence,

$$\|S_n^*f\|_p \leqslant C_p\|f\|_p, \quad 1 < p < \infty. \tag{3.11}$$

Of course, we may replace S_n^* by S_n in (3.11). The corresponding trigonometric result is due to M. Riesz [12]. Riesz proved that $\|\tilde{f}\|_p \leqslant C_p\|f\|_p$, $1 < p < \infty$, and used (3.3) to obtain the result for $S_n f$.

In his L^2 proof Carleson needs to use the fact that $\|\sup_{\epsilon>0}|\tilde{f}_\epsilon|\|_p \leqslant C_p\|f\|_p$ for some $p > 2$. (See [15; Vol. I, p. 279].) In the dyadic case, the analogue of the maximal conjugate function turns out to be $T_n^*f = \sup_k|S_{2^k}(S_n^*f)|$. Since $T_n^*f \leqslant \Lambda(S_n^*f)$, we can iterate previous results to obtain $\|T_n^*f\|_p \leqslant \|\Lambda(S_n^*f)\|_p \leqslant C_p\|S_n^*f\|_p \leqslant C_p'\|f\|_p$, $1 < p < \infty$. The same proof we gave which shows S_n^* is of weak type (1, 1) also shows that T_n^* is of weak type (1, 1). In the trigonometric case $\sup_{\epsilon>0}|\tilde{f}_\epsilon|$ is also related to the Hardy-Littlewood maximal function of $\tilde{f}$ and the same methods apply. (For example, see Cotlar [4].)

We have now completed the development of the results on which Carleson's proof is based, at least for the dyadic case. The next step is Carleson's proof itself or the Walsh function analogue found in [10]. Good luck!

APPENDIX. PROPERTIES OF THE WALSH FUNCTIONS

We will develop properties of the Walsh functions as functions of a real variable, even though the Walsh group 2^ω may be a more natural setting. From our point of view the dyadic rationals are somewhat exceptional because of the non-uniqueness of their (dyadic) decimal expansion. The countable collection of dyadic rationals will be excluded from our development.

ω will always denote a dyadic subinterval of [0, 1]. That is, $\omega = (k2^{-n}, (k+1)2^{-n})$, $n \geqslant 0$, $k = 0, \ldots, 2^n - 1$. Note that $\omega_1 \cap \omega_2 \neq \phi$ implies either $\omega_1 \subset \omega_2$ or $\omega_2 \subset \omega_1$.

Dyadic addition (taken from the Walsh group 2^ω) plays an important role in the development. For $x, t \in [0, 1]$, (x and t not dyadic rationals), write $x = \sum_{j=0}^{\infty}\eta_j 2^{-j-1}$ and $t = \sum_{j=0}^{\infty}\tau_j 2^{-j-1}$, where η_j and τ_j are either 0 or 1. The dyadic sum of x and t is $x \dotplus t = \sum_{j=0}^{\infty}|\eta_j - \tau_j|2^{-j-1}$. It is clear that $x \dotplus t = t \dotplus x$ and $x \dotplus t \epsilon [0, 1]$.

Two points $y = \sum_{j=0}^{\infty}\xi_j 2^{-j-1}$ and $y' = \sum_{i=0}^{\infty}\xi_j' 2^{-j-1}$, ξ_j, $\xi_j' = 0$ or 1, are in the same dyadic interval ω, $|\omega| = 2^{-N}$, if and only if $\xi_j = \xi_j'$, $j = 0, \ldots, N-1$. It follows that $x \dotplus \omega = \{x \dotplus t \colon t \epsilon \omega\}$ is a dyadic interval of the same length as ω. Since any Lebesgue measurable subset of [0, 1] can be approximated in measure by a finite union of disjoint dyadic intervals, we obtain that Lebesgue measure is translation invariant with respect to the dyadic addition $\dotplus$. Also (disregarding dyadic rationals) we have

$$x \dotplus t \in (0, 2^{-n}) \tag{A.1}$$

if and only if x and t belong to the same dyadic interval of length 2^{-n}.

The function r_0 is defined by $r_0(t) = 1$ if $0 < t < 1/2$, $r_0(t) = -1$ if $1/2 < t < 1$, and $r_0(t+1) = r_0(t)$. The nth Rademacher function is then $r_n(t) = r_0(2^n t)$, $n = 0, 1, \ldots$.

If

$$x = \sum_{j=0}^{\infty} \eta_j 2^{-j-1}, \ \eta_j = 0 \text{ or } 1,$$

(x not a dyadic rational) we have

$$\begin{aligned} r_n(x) &= r_0(2^n x) \\ &= r_0\left(\sum_{j=0}^{n-1} \eta_j 2^{-j-1+n} + \eta_n 2^{-1} + \sum_{j=n+1}^{\infty} \eta_j 2^{-j-1+n}\right). \end{aligned}$$

Since $\sum_{j=0}^{n-1}\eta_j 2^{-j-1+n}$ is an integer and $0 < \sum_{j=n+1}^{\infty}\eta_j 2^{-j-1+n} < 1/2$, we have $r_n(x) = (-1)^{\eta_n}$. Since η_n, $\tau_n = 0$ or 1 implies $(-1)^{|\eta_n - \tau_n|} = (-1)^{\eta_n}(-1)^{\tau_n}$, it follows that

$$r_n(x \dotplus t) = r_n(x) r_n(t). \tag{A.2}$$

$r_n(t)$ is constant as t varies over any ω with $|\omega| < 2^{-n}$. Also, $|\omega| \geqslant 2^{-n}$ implies $\int_\omega r_n(t)dt = 0$. Hence, we can divide (0, 1) into disjoint dyadic intervals of length 2^{-n_k} to obtain

$$0 \leqslant n_1 < n_2 < \cdots < n_k \text{ implies } \int_0^1 r_{n_1}(t) r_{n_2}(t) \cdots r_{n_k}(t) dt = 0. \tag{A.3}$$

For any nonnegative integer n, write $n = \Sigma_{j=0}^{\infty} \epsilon_j 2^j$, $\epsilon_j = 0$ or 1. The nth Walsh function is then $w_n(t) = \Pi_{j=0}^{\infty}[r_j(t)]^{\epsilon_j}$.

The product of any two Walsh functions is again a Walsh function. In particular, if $n = \Sigma_{j=0}^{\infty} \epsilon_j 2^j$ and $n' = \Sigma_{j=0}^{\infty} \epsilon'_j 2^j$, ϵ_j and $\epsilon'_j = 0$ or 1, let $n \dot{+} n' = \Sigma_{j=0}^{\infty} |\epsilon_j - \epsilon'_j| 2^j$. Then $w_n \cdot w_{n'} = w_{n \dot{+} n'}$. If $0 \leqslant s < 2^n$ and l is a nonnegative integer, $l2^n + s = l2^n \dot{+} s$. Hence, $w_{l2^n+s} = w_{l2^n} \cdot w_s$. $0 \leqslant s < 2^n$ implies that w_s is a product of Rademacher functions r_j with $j < n$, so w_s is constant on every ω with $|\omega| \leqslant 2^{-n}$. Also note that $w_{2^n} = r_n$.

It follows from (A.2) that $w_n(x \dot{+} t) = w_n(x) \cdot w_n(t)$.

Using (A.3), we see that

$$\int_0^1 w_n(t) w_m(t) dt = \begin{cases} 1, & \text{if } \quad n = m, \\ 0, & \text{if } \quad n \neq m. \end{cases}$$

Hence, $\{w_n\}_{n \geqslant 0}$ is an orthonormal sequence of functions on [0, 1].

For $f \in L^1[0, 1]$, the nth Walsh-Fourier coefficient of f is $c_n = c_n(f) = \int_0^1 f(t) w_n(t) dt$. The nth partial sum of the Walsh-Fourier series of f is

$$\begin{aligned} S_n f(x) &= \sum_{j=0}^{n-1} c_j w_j(x) \\ &= \sum_{j=0}^{n-1} \left[\int_0^1 f(t) w_j(t) dt \right] w_j(x) \\ &= \int_0^1 f(t) \left[\sum_{j=0}^{n-1} w_j(t) w_j(x) \right] dt \\ &= \int_0^1 f(t) \left[\sum_{j=0}^{n-1} w_j(t \dot{+} x) \right] dt \\ &= \int_0^1 f(t) D_n(t \dot{+} x) dt, \end{aligned}$$

where $D_n(t) = \Sigma_{j=0}^{n-1} w_j(t)$ is the nth Dirichlet kernel.

It is useful to note

$$\sum_{j=2^k}^{2^{k+1}-1} w_j = \sum_{j=0}^{2^k-1} w_{2^k+j} = \sum_{j=0}^{2^k-1} w_{2^k} \cdot w_j = r_k \sum_{j=0}^{2^k-1} w_j. \tag{A.4}$$

A finite induction argument which uses (A.4) shows

$$\sum_{j=0}^{2^k-1} w_j(t) = \prod_{\nu=0}^{k-1} (1 + r_\nu(t))$$

$$= \begin{cases} 2^k, & \text{if } \; 0 < t < 2^{-k}, \\ 0, & \text{if } \; 2^{-k} < t < 1. \end{cases} \tag{A.5}$$

We can combine (A.1) and (A.5) to obtain

$$S_{2^N} f(x) = \int_0^1 f(t) D_{2^N}(t \dotplus x)\, dt$$

$$= \frac{1}{|\omega|} \int_\omega f(t)\, dt,$$

where $x \in \omega$, $|\omega| = 2^{-N}$. One can then show that $\lim_{N\to\infty} S_{2^N} f(x) = f(x)$ a.e. (For example, see Section 2.) Thus, if $c_n(f) = 0$ for all n, then $S_{2^N} f(x) \equiv 0$, so $f = 0$ a.e. In particular, this shows $\{w_n\}_{n \geqslant 0}$ is a complete orthonormal sequence, so we have Parseval's equality,

$$\sum_{n=0}^{\infty} |c_n|^2 = \int_0^1 |f(t)|^2 dt.$$

The modified Dirichlet kernel is $D_n^*(t) = w_n(t) D_n(t)$. We will need

$$n = \sum_{j=0}^{\infty} \epsilon_j 2^j, \; \epsilon_j = 0 \text{ or } 1, \text{ implies} \tag{A.6}$$

$$D_n^*(t) = \sum_{j=0}^{\infty} \epsilon_j \delta_j^*(t),$$

where

$$\delta_j^*(t) = \sum_{\nu=2^j}^{2^{j+1}-1} w_\nu(t).$$

Note that (A.6) is clear from (A.4) if $n = 2^N$. This fact can be used in a finite induction argument to prove (A.6) for all n. An alternate proof is obtained by noting that D_n^* is exactly the sum of the Walsh functions $w_{n \dot{+} \mu}$, $\mu = 0, \ldots, n-1$. If $\epsilon_j = 1$, for each term w_ν $(2^j \leqslant \nu < 2^{j+1})$ of δ_j^*, we have $\nu = n \dot{+} \mu$, where $\mu = n \dot{+} \nu$ satisfies $0 \leqslant \mu < n$. (A.6) follows, since $j \to n \dot{+} j$ is a one-to-one mapping of the set of nonnegative integers onto itself.

From (A.4) and (A.5) we see that $\delta_j^*(t)$ is constant on any ω which does not contain $(0, 2^{-j})$. Using (A.1), we then have that $\delta_j^*(t \dot{+} x)$ is constant as t varies over any ω which does not contain x. From (A.6) we then conclude that $D_n^*(x \dot{+} t)$ is constant as t varies over any ω which does not contain x.

REFERENCES

1. Besicovitch, A. S., "A general form of the covering principle and relative differentiation of additive functions II," *Proc. Cambridge Philos. Soc.*, **42** (1946), 1–10.
2. Calderón, A. P., and A. Zygmund, "On the existence of certain singular integrals," *Acta Math.*, **88** (1952), 85–139.
3. Carleson, L., "On convergence and growth of partial sums of Fourier series," *Acta Math.*, **116** (1966), 135–157.
4. Cotlar, M., "Some generalizations of the Hardy-Littlewood maximal theorem," *Rev. Mat. Cuyana*, **1** (1955), 85–104.
5. Fine, N. J., "On the Walsh functions," *Trans. Amer. Math. Soc.*, **65** (1949), 372–414.
6. Garsia, A. M., *Topics in Almost Everywhere Convergence*, Markham, Chicago, 1970.
7. Hardy, G. H., and J. E. Littlewood, "A maximal theorem with function-theoretic applications," *Acta Math.*, **54** (1930), 81–116.
8. Hunt, R. A., "Comments on Lusin's conjecture and Carleson's proof for L^2 Fourier series," P. L. Butzer and B. Sz.-Nagy, eds., *Linear Operators and Approximation II*, (ISNM 25) Birkhäuser Verlag, Basel/Stuttgart, 1974, 235-245.
9. Hunt, R. A., "On the convergence of Fourier series," D. T. Haimo, ed., *Orthogonal Expansions and their Continuous Analogues*, Southern Illinois University Press, Carbondale, 1968, 235–255.

10. Hunt, R. A., "Almost everywhere convergence of Walsh-Fourier series of L^2 functions," *Actes, Congrès Intern. Math.*, **2** (1970), 655–661.

11. Lusin, N., "Sur la convergence des séries trigonométriques de Fourier," *C. R. Acad. Sci. Paris*, **156** (1913), 1655–1658.

12. Riesz, M., "Sur les fonctions conjuguées," *Math. Z.*, **27** (1928), 218–244.

13. Stein, E. M., *Singular Integrals and Differentiability Properties of Functions*, (Princeton Mathematical Series No. 30) Princeton University Press, Princeton, 1970.

14. Stein, E. M., "On limits of sequences of operators," *Ann. of Math.*, **74** (1961), 140–170.

15. Zygmund, A., *Trigonometric Series*, 2nd. ed., Cambridge University Press, New York, 1959.

HARMONIC ANALYSIS AND H^p SPACES*

Charles Fefferman

1. CLASSICAL THEORIES

It seems to be the lot of the speakers at this conference to give expository talks to an audience containing a tremendous number of experts, so I'm afraid a fair number of people are going to hear me explaining their theorems to them. But that will just have to be. Now the H^p spaces arose originally from some simple problems of function theory but soon became very closely bound up with Fourier analysis. That was very natural simply because the methods in use at the time were the so-called complex methods in which you study Fourier series by associating analytic functions to them. Therefore it was natural to expect that more powerful theorems could be proved for H^p than for L^p. Since this is an expository article I'll define H^p.

I'll start on the circle. We have an analytic function $F(z)$.

Given a circle centered about the origin with the radius a little

*This paper was taken from the actual videotape of the conference. It could not have been done without the help of Marshall Ash.

less than 1, you form the integral of the p-th power of the absolute value.

If the sup of

$$\left(\frac{1}{2\pi} \int_0^{2\pi} |F(re^{i\theta})|^p \, d\theta \right)^{1/p}$$

over all $0 < r < 1$ is finite, then F is said to belong to H^p.

The connection with Fourier series comes instantly because if you have a function expanded in a Fourier series

$$f(\theta) = \sum_{-\infty}^{\infty} a_k e^{ik\theta},$$

then if you put $z = e^{i\theta}$ and think of z as a point either on the unit circle or inside, then of course this is the sum $\Sigma a_k z^k$. If you should be so lucky that all the a_k are zero on one side, then you have a sum from only 0 to ∞, a power series, and then f is simply an analytic function.

So perhaps a Fourier series containing only terms with positive frequencies is a lot better than an arbitrary Fourier series and perhaps the techniques of the proofs come from complex variables. This is the original point of view in the subject. Everything has since become very much generalized and turned inside out. You'll see what becomes of it later on, but for the time being, complex analysis is what the subject is all about.

Well, this is the subject. What are the theorems? I have a list of some theorems here. To begin with, there are certain elementary things about the existence of boundary values. I won't talk about difficult theorems—just easy ones. F is going to be in H^p, p is going to be anything—it doesn't much matter—let's say $\infty > p > 0$: I won't talk about H^∞.

(a) *At almost every point of the circle* $[0, 2\pi)$, $F(z)$ *has a non-tangential limit.*

This means that if you take a point on the unit circle and draw a little triangle, then the limit of $F(z)$ as z approaches $e^{i\theta}$, always staying within the triangle, exists. Let's give the triangle a name for all eternity—$\Gamma(\theta)$. (See Figure 1.)

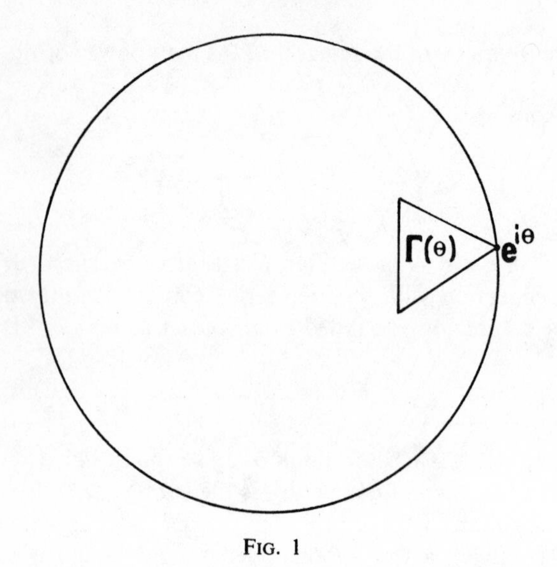

FIG. 1

Maybe I should say right at the beginning that I'll switch back and forth completely at random between the unit disc and the upper half plane.

Instead of integrating analytic functions over circles, you integrate them over lines which are raised some small height (Figure 2). Here, instead of $\Gamma(\theta)$, I'll work with $\Gamma(x)$ which is just another small triangle—well, perhaps not so small—we'll take the whole crosshatched region.

There is a sharper version of theorem (a). It is the so-called maximal theorem which says that not only the limit, but even the sup—$F^*(\theta) = \sup_{z \in \Gamma(\theta)} |F(z)|$—exists as an L^p function.

(b) $F^*(\theta) \in L^p$.

Now these limits don't merely exist—they have some properties. For example,

(c) *$F(\theta)$ is non-zero almost everywhere*,

and you can say a little bit more. Let me quote for you the F. and M. Riesz theorem. This familiar theorem says,

(d) *Suppose $d\mu$ is a measure of finite mass, with a Fourier*

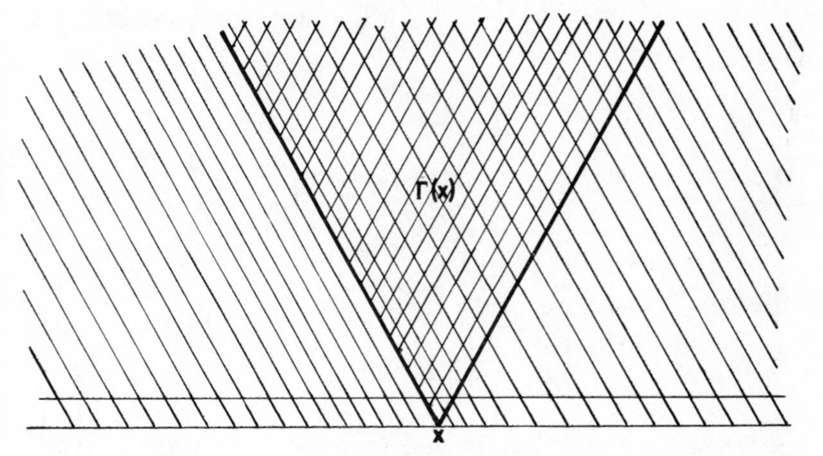

FIG. 2

expansion containing only positive frequencies

$$d\mu \sim \sum_{k=0}^{\infty} a_k e^{ik\theta}.$$

Then actually $d\mu = f d\theta$ with $f \in L^1$, that is, the measure is absolutely continuous.

Although theorems (a)–(d) deal only with function theory, H^p spaces also have important Fourier-analytic properties. To illustrate, I will throw in one theorem, chosen at random:

(e) *If a power series $f = \sum_{k=0}^{\infty} a_k e^{ik\theta}$ is of bounded variation on the circle $[0, 2\pi)$, then it is absolutely convergent.*

These are standard theorems. I want to tell you a little bit about the proofs so that you get the idea—the flavor—of the subject as it existed, let's say before 1935 or so.

The basic plan in proving (a)–(e) and such things is to use complex function theory. Now it turns out that the most important step is to get your hands on the set of zeros. It turns out that all the theorems are very nice and easy for functions that don't have any zeros. The nasty problems arise from zeros—the difficulty in the subject is to get rid of them. Why? Let me just say something about that. Let's look, for example, at (a). In the simplest case, let's say p is 2. We have then a Fourier series of power series type

$$f(\theta) \sim \sum_{k=0}^{\infty} a_k e^{ik\theta} \in L^2.$$

The corresponding power series is the sum

$$\sum_{k=0}^{\infty} a_k z^k.$$

If we put $z = re^{i\theta}$, then we've got

$$\sum_{k=0}^{\infty} a_k r^k e^{ik\theta},$$

and we'd like to say that for each fixed $r < 1$ this function of θ belongs to L^2 with norm independent of r. Of course that simply means that the Fourier series represents an L^2 function on the boundary—the sum of the squares of the coefficients is convergent—and so this is a very nice function. Now we can be very crude. You don't need to use the fact that this is a power series. You simply have an L^2 function. You compute its Poisson integral; that is, given the function f, you extend it to be harmonic in the unit disc. My assertion is just a very elementary theorem of the convergence of the Poisson integral to boundary values.

Given f, we have u = P.I.(f) = the Poisson integral of f given by

$$u(r, \theta) = P_r * f(\theta) = \frac{1}{2\pi} \int_0^{\pi} f(\theta - t) \frac{1 - r^2}{1 - 2r \cos t + r^2} \, dt.$$

The basic thing about the Poisson kernel is that it looks like Figure 3. It's a nice positive curve with area 1—a nice approximate identity—so the fact that the limit as $r \to 1$ of $\Sigma a_k r^k e^{ik\theta}$ is $f(\theta)$ almost everywhere is a simple consequence of the Lebesgue theorem on the differentiation of the integral. Where does that work? That works for $p \geqslant 1$, so theorem (a) is trivial for $p \geqslant 1$. Similarly, (b) is the standard theorem of Hardy and Littlewood for $p > 1$; but already if $p = 1$, this idea flops—at least the trivial proof that

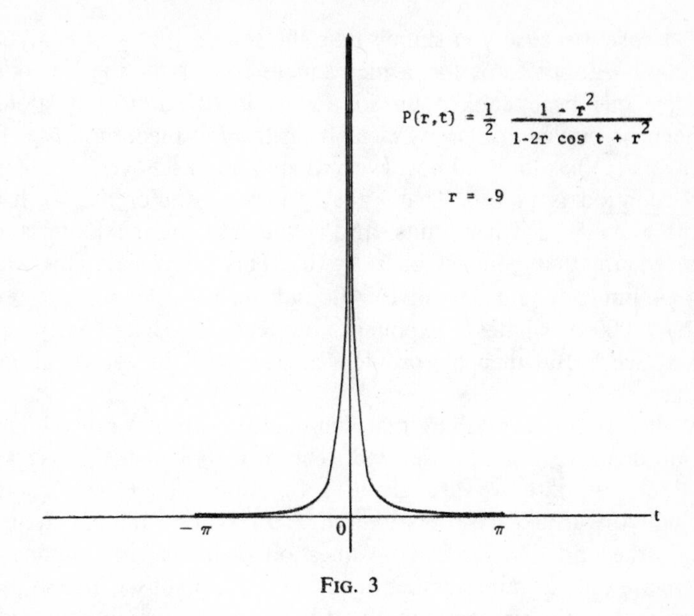

FIG. 3

you would like to use won't work. Of course there is no hope of obtaining (c) without using analyticity. There are all sorts of functions which vanish somewhere but not everywhere. Similarly for (d) and (e).

That's about as much as you can do without the analyticity hypothesis. Now you have to bring analyticity into play somehow. Bear in mind that you are not merely dealing with any old harmonic function—the Poisson integral of any old thing—but in fact you have a power series and therefore an analytic function.

Let's look for example at the maximal theorem and the existence of nontangential limits—they are practically the same thing. You have an analytic function $F(z)$—it lives in the disc—on circles it has bounded L^p norm (say $\frac{1}{2} < p < 1$) and the hope is that at almost every point we have $\lim_{\substack{z \in \Gamma(\theta) \\ z \to \theta}} F(z)$ existing. Let's suppose for a minute that F has no zeros. If F has no zeros, then the problem is completely trivial; that is, it is reducible to the

$p > 1$ case, because you simply take the square root—an analytic function without zeros has a nice square root. Now the function $F^{1/2}(z)$ has been cooked up so that it is L^{2p} uniformly along concentric circles. It's pretty clear from the L^{2p} theorem $(2p > 1)$ that $F^{1/2}(z)$ has limits almost everywhere and, moreover, its maximal function is in L^{2p}. That's all there is to it. For if $F^{1/2}$ has limits, then F itself has limits; and if you take the maximum and then square that you get back to (b). This is typical. This was $p > \frac{1}{2}$, but p could be anything, and instead of $\frac{1}{2}$, you use 1/2,000,000 or whatever exponent you need in order to get yourself above 1. But then the problem arises: what do you do about the zeros?

Well, virtually everything that you could possibly want to know about the zeros can be decided. The key tool is the Blaschke product. Suppose we are given a function $F(z) \in H^p$. (Let's exclude the trivial case. Assume $F \neq 0$.) Where are F's zeros? Inside the unit disc there is no question of having lots of them. There are only a finite number inside a compact subset, and so the worst that could happen is that you have some countable number of zeros with perhaps enormous multiplicities, which tend to the boundary, we don't know how fast. For an arbitrary analytic function, that's about all you can say. But since F is somewhat reasonable, in fact $F \in H^p$, some restrictions are imposed on the zeros. Let's say the zeros can be listed $a_1, a_2, \ldots, a_k, \ldots$. They tend to the boundary, and I list them according to multiplicity so that a single number a may show up lots of times. The basic fact about these zeros is that the sum of the distances to the boundary converges. Conversely, given a set of $\{a_k\}$ that satisfy $\Sigma(1 - |a_k|) < \infty$, you can cook up a very nice function which has exactly these $\{a_k\}$ as its zeros. Let me tell you what it is. For each fixed a_k you can write down the linear fractional transformation which preserves the unit disc, but takes the point a_k to the origin

$$b_k(z) = \frac{|a_k|}{-a_k} \cdot \frac{z - a_k}{1 - \bar{a}_k z}.$$

The normalizing constant has absolute value 1—it's fudged so that if you put $z = 0$, b_k becomes real, in fact positive. Then b_k is a function with a zero of exactly the right sort and its absolute value is 1 on the boundary

$$|b_k(e^{i\theta})| = \left| \frac{e^{i\theta} - a_k}{1 - \bar{a}_k e^{i\theta}} \right|$$

$$= \left| \frac{e^{i\theta} - a_k}{\overline{(e^{i\theta} - a_k)}\, e^{i\theta}} \right| = 1.$$

In particular, I can take the product of all of them. With any luck, that will converge. Let me give it a name:

$$B(z) = \prod_{k=1}^{\infty} b_k(z).$$

If we are given any sequence of numbers in the disc $\{a_k\}$ satisfying $\Sigma(1 - |a_k|) < \infty$, then it follows that actually the product defining B converges uniformly in compact subsets of the unit disc, so that $B(z)$ is well defined, analytic, and has boundary values of absolute value 1 almost everywhere on the edge [9, vol I, p. 274]. Once you've got B, the hope is you can divide out by it—B will capture all the zeros and what's left will be some function that controls the size of F, but has no zeros. If $F \in H^p$, you can write $F(z) = G(z)B(z)$, where G is some nice function. The Blaschke product B carries the zeros, but on the other hand in terms of its size it's very nice—it's of absolute value 1 on the boundary. The function G has no zeros, but has the same H^p norm as F. (Formally, B is of absolute value 1 on the boundary, F is L^p on the boundary, and hence $G = F/B$ has to be L^p on the boundary. I won't give the details here [9, vol I, p. 274].)

Once this theorem is known, then lots of things are very easy. For example, let's prove the maximal theorem. It would be nice to take the square root of $F \in H^p$, but we've seen that we can't do that. On the other hand, you can write $F = GB$. The function B is the Blaschke product; that's already nice. So let's write $F = (G^{1/2})(G^{1/2}B)$. Since G has no zeros, it has a nice square root. Both $G^{1/2}$

and $G^{1/2}B$ are in H^{2p}. In other words, while you can't simply take the square root, still the idea goes through. You can express F as the product of two things in H^{2p}. Similarly, if I had to express F as a product of 50,000 things, I would take all of them except the last one as appropriate roots of the function G and then I'd tack on B in the last factor at the end. So if you believe the business about Blaschke products, that proves (b) for all $p > 0$. Once that's true, the F. and M. Riesz theorem (d) is a triviality because you look at the function $F(re^{i\theta})$ and you see what happens as r tends to 1. According to (a), it tends to a limit almost everywhere. But more than that, (b) tells you that the convergence is dominated. Then the good old Lebesgue-dominated convergence theorem tells you that then the convergence is of the nicest possible sort: in L^1 norm to a nice L^1 function, and (d) is proved. Theorem (c) doesn't follow from the facts I've stated about Blaschke products, but it does follow from their proofs. I will omit the proof of (e) [9, vol. I, p. 285]. Let me say just a little about Blaschke products and the way you prove things about them. Everything comes down to a very simple formula. Let's take only a very nice function with a finite number of zeros. Then one has Jensen's formula:

$$\log|F(0)| - \sum_k \log|a_k| = \frac{1}{2\pi}\int_0^{2\pi} \log|F(e^{i\theta})|\,d\theta.$$

(Both here, and in the discussion which follows, we will assume, just to ease notation, that $F(0) \neq 0$.)

Why? It's very easy. If F has no zeros, then $\log(F(z))$ is analytic, $\log|F(z)|$ being the real part is harmonic, and therefore the value at the center is the average of the values on the circle, and the term $\Sigma\log|a_k|$ doesn't appear. If F has zeros, the term $\Sigma\log|a_k|$ does appear. Then you write down the formula for a function with no zeros—namely F divided by its Blaschke product. There is no problem because the Blaschke product is finite; and $\Sigma\log|a_k|$ is the extra term you get from the Blaschke product. So Jensen's formula holds for functions with finitely many zeros.

Now, by being careful, one can prove that Jensen's formula actually holds for functions with an infinite number of zeros.

That's perhaps not such a meaningful statement; after all I'm trying to prove that boundary values exist. What it means is that any sort of conclusions about the zeros that you can derive from this formula formally can actually be seen by doing everything carefully.

We don't know that $I = \int_0^{2\pi} \log|F(e^{i\theta})|d\theta$ converges, because if, for example F were 0 on a set of positive measure, then I would be $-\infty$. This is not very good. On the other hand, you do know at least that $\int_0^{2\pi} \log^+ |F(e^{i\theta})|d\theta$ converges—the integrand is dominated by $|F(e^{i\theta})|^p$ for any $p > 0$. So we know that I is either finite or $-\infty$. On the other hand, $\log|F(0)|$ is just some finite number. What about $\Sigma\log|a_k|$? Its convergence is equivalent to that of $\Sigma(1 - |a_k|)$ (since all but finitely many a_k satisfy $2/3 < |a_k| < 1$ and, by Taylor's formula, if $t \in [2/3, 1)$, $|[\log t - (t-1)]/(t-1)| \leqslant 3/8$, so that $5/8 < (\log t)/(t-1) < 11/8$). *A priori*, we don't know if $\Sigma\log|a_k|$ converges. We hope it does. What we know about $-\Sigma\log|a_k|$ is that it is finite, or $+\infty$. From Jensen's formula, we see that neither of the infinities can occur, because they don't match $(+\infty \neq -\infty)$.

This should give the idea how proofs with Blaschke products go. I shouldn't say too much more about it. But I do want to speak about one other class of theorems—these are the theorems of Littlewood-Paley theory. Before I leave complex methods, I just want to point out that there is a class of deeper theorems for the H^p classes.

Littlewood-Paley theory deals with problems something like this. Suppose that you take a Fourier series, $f(\theta) = \Sigma_{k=-\infty}^{\infty} a_k e^{ik\theta}$. You would like to know whether $f(\theta)$ is in L^p. This is a well-known trivial thing if p is 2. But if p isn't 2, it's much nastier. There is no way of knowing very accurately or of writing down a good answer. Part of the reason that's so hard is that whereas in the L^2 case all that's important is the size of the coefficients, now things are quite different. For p different from 2, the arguments of the coefficients are extremely important in determing the size of the function. A typical problem that Littlewood-Paley theory deals with is: Suppose you stick in some plus and minus signs, getting

$$\sum_{k=-\infty}^{\infty} \epsilon_k a_k e^{ik\theta}$$

(each $\epsilon_k = +1$ or -1). You would like to know conditions under which this new series is still the Fourier series of a function which is still in L^p. If you can get a handle on this problem, it tells you something about how hard it is to pin down the size of a Fourier series. That is, suppose that every choice of plus and minus (i.e., of sequence $\{\epsilon_k\}$) were all right. That would tell you that only the sizes were important (which is admittedly false). On the other hand, if it turned out that no $\{\epsilon_k\}$ worked, that would tell you that the problem was so extremely delicate that there could be no hope of any progress. The truth is somewhere in between—perhaps closer to the second. A typical theorem of Littlewood-Paley theory which is a consequence of a result of Marcinkiewicz—a special case of the so-called Marcinkiewicz multiplier theorem—is this:

Suppose that in each dyadic block of indices the sign of the multiplier is constant (that is, ϵ_k is constant for $2^N \leqslant k < 2^{N+1}$). So you cut the series into dyadic blocks (also for negative frequencies) and then change each dyadic block. Either keep it the same or else replace it by minus itself. The theorem is that this process is O.K. That is, if you started in L^p, you end up in L^p, for $1 < p < \infty$.

This is the sort of thing one is aiming for. I'd like to say a few words about some of the tools that are used in proving this, and I'd just like to point out that they have something to do with H^p.

For simplicity let f be of power series type $f = \Sigma_{N=0}^{\infty} f_N$, where $f_N(\theta) = \Sigma_{k=2^N}^{2^{N+1}-1} a_k e^{ik\theta}$ is the sum of those terms in the Fourier series arising in a dyadic block. Then the basic thing to prove is the set of inequalities that the L^p norm of f is more or less equivalent to the L^p norm of the square root of the sum of the squares of the f_N.

$$\|f\|_p \sim \left\| \left(\Sigma |f_N|^2 \right)^{1/2} \right\|_p ,$$

where $A \sim B$ means A/B and B/A are bounded. There's no point in getting into the proof of this here but it's clear that once such a thing is true, then once you multiply the f_N's by plus or minus ones, any way you like, then the right hand side is not affected at all. The way that you control the relation

$$\|f\|_p \sim \left\| \left(\sum |f_N|^2 \right)^{1/2} \right\|_p,$$

is by bridging the gap between the two functions by some compromise which still looks recognizably like both of them, and involves complex function theory, so that the complex method can be applied. This is the area function

$$S(f)(\theta) = \left(\int \int_{\Gamma(\theta)} |f'(z)|^2 dz \, \overline{dz} \right)^{1/2}$$

where $\Gamma(\theta)$ is the non-tangential region and $f(z)$ is the analytic extension of $f(\theta)$ to the interior of the disc [1], [9, vol II, p. 207]. Aside from the square root, this has a natural interpretation as the area of the image of the non-tangential region under the mapping f (which doesn't help you, by the way). What you prove then is that $\|f\|_p \sim \|S(f)\|_p$ and $\|S(f)\|_p \sim \|(\Sigma |f_N|^2)^{1/2}\|_p$. And so it came about that S and a few of its variants come into prominence in proving Marcinkiewicz's theorem and related results. In a while we'll see that actually these functions have very great heuristic significance. They play a big role as a tool in dealing with the H^p spaces; actually the business about the area is quite superficial—no pun intended. In fact, the connection is quite different and has to do with gambling. In any event, the one theorem I want to quote having to do with H^p spaces is:

THEOREM: *The analytic function $F(z) \in H^p$ if and only if $S(F) \in L^p$, $0 < p < \infty$.*

This is in the classical situation. By the way, this equivalence has found surprising applications in the study of the commutator integral of Calderón, a deep problem seemingly having nothing to do with the S-function. Calderón showed that using this fact you could control the commutator integral, and it's only just very recently that Coifman and Meyer have shown that indeed commutators have nothing to do with S-functions [2], [4]. That's all I'd like to say about the classical theory for the time being, so now let's start all over again.

Let's work in Euclidean n-space R^n. It happens that functions of several complex variables are entirely the wrong way to generalize H^p to R^n. After all, the zeros of functions on the disc are a simple-minded little set of points, but the zeros of a function of several complex variables are something much hairier—they are a variety which degenerates and proliferates and does very strange things as you approach the boundary. So there's very little hope of proving any good theorems along these lines.

So instead of analytic functions, let's look at harmonic functions in Euclidean n-space. Now hopefully, harmonic functions have intimate connections with real variables. In particular in Euclidean space, you have the theory of singular integrals. Back in the half plane, you have the very famous fact that, given an analytic function $u(z) + iv(z) = F(z)$ on the boundary, there is an operator which links u to v, an operator of tremendous importance in Fourier analysis—the Hilbert transform. In fact, v is given by the principal value integral

$$v(x) = \frac{1}{\pi}\int_{-\infty}^{\infty} \frac{u(y)dy}{x-y}. \qquad \text{(We say } v = \tilde{u}.\text{)}$$

In n dimensions, there are basic generalizations of this operator—the singular integrals. The hope is going to be that there will be some theory of H^p spaces in which these generalizations of the Hilbert transform will play some role in connecting one component to another. The natural generalizations are the singular integrals, that is operators of the form

$$Tf(x) = \int_{R^n} \frac{\Omega(y/|y|)}{|y|^n} f(x-y)dy$$

where Ω has mean value zero $\left(\int_{|t|=1} \Omega(t)dt = 0\right)$ and is not too big, say, $\int_{|t|=1} |\Omega(t)|^2 dt < \infty$, and integration is in the principal value sense $\left(\int_{R^n} = \lim_{\epsilon \to 0} \int_{|y|>\epsilon}\right)$.

Particular cases are the Riesz transforms

$$R_j f(x) = c_n \int_{R^n} \frac{y_j}{|y|^{n+1}} f(x-y)dy, \quad j = 1, 2, \ldots, n.$$

These Riesz transforms come up, for example, when you begin to study the operator $\partial/\partial x_j$—partial differentiation. In order to make it an operator of degree 0, you divide by the square root of the Laplacian, and the resulting operator, formally $(\partial/\partial x_j)(-\Delta)^{-1/2}$, is, up to a constant, the j-th Riesz transform R_j. Now j runs from 1 to n, so instead of one Hilbert transform, we've got n of them. These play the role analogous to the Hilbert transform if you use a certain generalization of analytic function. What's going to be an "analytic function"? The right way to think of an analytic function is that it's a system (u, v) which satisfies the Cauchy-Riemann equations. From this point of view, the Cauchy-Riemann equations tell you that this system (u, v) forms the gradient of a harmonic function. So we'd like to regard R^n as the boundary of some bigger thing, say the upper half of R^{n+1} : $R^{n+1}_+ = \{x \in R^{n+1} : x_0 > 0\}$. We look then at a system

$$(u_0, u_1, \ldots, u_n)$$

of $n + 1$ functions which arises as the gradient of a big harmonic function, that is,

$$u_j = \frac{\partial U}{\partial x_j},$$

where U is harmonic. What are the equations relating the u_j's? The natural consistency conditions are simply

$$(1)\quad \frac{\partial u_i}{\partial x_j} = \frac{\partial u_j}{\partial x_i} \quad \text{and} \quad (2)\quad \sum_{i=0}^{n} \frac{\partial u_i}{\partial x_i} = 0.$$

Equations (1) say that $(u_0, \ldots, u_n)$ are the gradient of something, and equation (2) says that the something is harmonic. Stein and Weiss had the idea of making an H^p theory out of this [8]. That is, you consider systems of harmonic functions of this kind with the property that if you take the boundary (R^n), elevate it slightly, and compute the integral of the p-th power of the square root of the sum of the squares of the u_j's,

$$\left(\int_{R^n} (u_0^2 + u_1^2 + \cdots + u_n^2)^{p/2} (h, x_1, \ldots, x_n) dx \right)^{1/p}$$

then this expression should be uniformly bounded as the height $h = x_0$ of the translate of R^n being integrated over tends to 0. By the way, what is the meaning of x_0? The coordinates x_1, $x_2, \ldots, x_n$ define the horizontal plane, but x_0 is the coordinate for up and down. There is no best choice of coordinates $(x_1, \ldots, x_n)$; the $u_1, u_2, \ldots, u_n$ can be rotated—you can think of them as a vector; but the direction "up" is very natural, so u_0 is singled out. There is a theory of H^p spaces for these, in which the Riesz transforms play the role of the Hilbert transform. In particular, if you restrict the system $(u_0, u_1, \ldots, u_n)$ from R^{n+1}_+ to the boundary R^n, then you can get the $u_i|_{R^n} (i \geqslant 1)$ as Riesz transforms of $u_0|_{R^n}$. Lots of the properties (existence of boundary values, the F. and M. Riesz theorem, and so forth) go through in this generality. Let me mention a couple of things that don't go through—at least not obviously. First of all, what about the question of vanishing? Suppose that you have a system $(u_0, u_1, \ldots, u_n)$ which is very nice, but which is zero in a set of positive measure on the boundary. Is it zero? Nobody knows. People have worked on it for a long time, but I think it's fair to say that there's no progress. In fact, it's rather a remarkable accident that anything at all can be done with these systems. I'd like to show you the miracle that makes it possible to squeeze something out of this definition of H^p.

Consider the size of the system—$(\Sigma_k |u_k(x)|^2)^{1/2}$. This is the norm of a certain vector-valued function of x—and of course $(u_0, u_1, \ldots, u_n)$ is a vector of harmonic functions, and so it's obvious that $(\Sigma |u_k|^2)^{1/2}$ is subharmonic, (i.e., has non-negative Laplacian). In one complex variable, what we used was that $\log |F(z)|$, which is a lot worse than just F itself, is in fact subharmonic—in fact it's exactly harmonic except at certain awful points. Now in R^n, strangely enough, the norm raised to the p-th power, $(\Sigma |u_k|^2)^{p/2}$, is subharmonic for certain positive p smaller than 1. From that fact you can reduce matters from p smaller than 1 to p bigger than 1, more or less as in the proof of (a) above. However, the Stein-Weiss H^p theory is technically a lot harder than the classical theory. Back in one complex variable, we had available the whole arsenal of complex analysis (e.g., Cauchy's integral formula and residues, conformal mapping, Blaschke products); now in R^n there is only the single weapon of subharmonicity of $(\Sigma_k |u_k|^2)^{p/2}$ to prove our theorems for us. (I'm oversimplifying slightly.) So to prove analogues of familiar theorems, you have to work much harder. It's a remarkable accident that it works at all.

One very nice property of the logarithm is that it becomes $-\infty$ at zero; and therefore, in the one-dimensional case, from the fact that you had the logarithm controlled, you knew something about the zeros, and now that no longer works. There's one other strange thing—that $\Delta(\Sigma |u_k|^2)^{p/2} \geqslant 0$ doesn't work for all p, but only for $p > (n-1)/n$, and as a result, the theory looks very strange. You only get theorems about H^p for $p > (n-1)/n$. You can get some sort of theory for p smaller—this is the discovery of Calderón and Zygmund; but then you have to change the whole definition [8]. Instead of looking at systems $(u_0, u_1, \ldots, u_n)$ which are single gradients of harmonic functions, you have to look at big tensors of harmonic functions—$u_{i_1, i_2, \ldots, i_k}$—which satisfy consistency conditions so that they are the k-th gradient of a big harmonic function U:

$$\frac{\partial^k U}{\partial x_{i_1} \cdots \partial x_{i_k}} = u_{i_1, i_2, \ldots, i_k}.$$

And so the very definition (the set of Cauchy-Riemann equations) changes as p changes. One other problem about which there is virtually no information is whether this machinery is really necessary. It's certainly clear from looking at easy examples that this trick of going from $p < 1$ to $p > 1$ fails below $p = (n-1)/n$; but that doesn't mean that the theorems you want to prove are false. Let me take this for granted, and just say a little bit about how you can prove some theorems about H^p in R^n. I'll mention the proofs of two results. The first is that if you have a Cauchy-Riemann system $F = (u_0, u_1, \ldots, u_n)$ in H^p, p bigger than the critical value $(n-1)/n$, then F has non-tangential limits almost everywhere and the maximal function F^* (which, analogous to the one-dimensional maximal function, is the sup over a cone) belongs to L^p. Let me talk first about the maximal function F^*. Let's say for convenience F is in H^1. Already trivial methods work for F in H^p if $p > 1$. Consider $|F|^r$. If r is just a little bit less than 1, $|F|^r = (\Sigma |u_k|^2)^{r/2}$ will be subharmonic. That means that if I look at the boundary values of $|F|^r$, say $g = |F|^r$ restricted to R^n (R^n = the boundary of R^{n+1}_+), and if I then take the Poisson integral of g, then $|F|^r \leqslant$ P.I.(g). This is the property of subharmonic functions, that they are dominated by harmonic functions with the same boundary values. On the other hand, $g \in L^{1/r}$, where mercifully $1/r$ is *bigger* than one. Therefore we're in the situation in which the maximal function works for basically trivial reasons. (Trivial means that you don't use the Cauchy-Riemann equation—you just use facts about the differentiation of the integral.) So you know that the maximal function of the Poisson integral of g belongs to $L^{1/r}$. Since $|F|^r$ is dominated by P.I.(g), that's all you need. But notice that you've lost something. You know also that this Poisson integral has limits almost everywhere. But you can't deduce the theorem about F having non-tangential limits from that. For, unfortunately, now you know only that $|F|^r$ is dominated by something; you don't know that F tends to a limit just from looking at this proof, and therefore you need an extra idea. That extra idea is a theorem of Calderón. The theorem says that if u is any harmonic function defined in the upper space R_+^{n+1} and you consider the two subsets of the boundary R^n defined by $E_1 = \{x | u \text{ is bounded in } \Gamma(x)\}$ and $E_2 = \{x | u \text{ has a limit as } z \text{ tends to } x, \text{ staying in } \Gamma(x)\}$ (where $\Gamma(x)$ is any cone with tip at x and pointing upwards—see Figure 4); then, up to a set of measure 0, these two sets are the same [9, vol II, p. 323]. Since we already know the maximal theorem, it follows that E_1 is almost all of R^n. From Calderón's theorem E_2 is also almost everything.†

2. MODERN THEORIES

I'd like to start off with the theorem of Calderón. Suppose u is a harmonic function on R_+^{n+1} and E is a set of positive measure in R^n such that at every point x of E, there is a cone $\Gamma(x)$ such that $\sup_{z \in \Gamma(x)} |u(z)|$ is finite. Then at almost every point of E, u has a non-tangential limit.

Let's look at this even in one dimension. The basic idea is this. You've got here a rotten harmonic function defined in a very nice domain. The idea is to replace the problem by another one involving a very nice harmonic function on a rotten domain. The way you construct the domain is as follows. Let's say that the set E is closed—that's no loss of generality. So the complement is a bunch of intervals— as shown in Figure 4.

We know that for each point x of the set E, the harmonic function $u(x)$ remains bounded in the whole sector $\Gamma(x)$. Let's say that $u(x)$ remains bounded by the same constant no matter which point x of E that you take. (That's just a small technical reduction that can be done easily by making the set E a little smaller—you sacrifice a set of measure ϵ and since ϵ is arbitrarily small, it doesn't matter.) So let's say, for example, that u is

---

†It might be thought that this trespasses illegally on the territory of Burkholder, but in a smoke-filled room an agreement was worked out by which it doesn't. And so, more next time on probability.

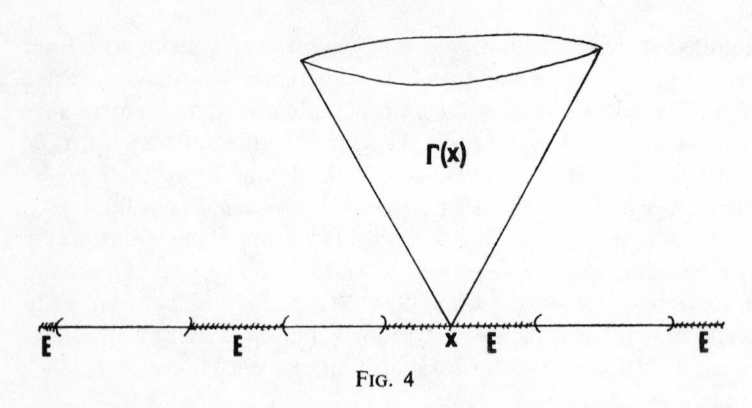

FIG. 4

dominated by 1 throughout the sector $\Gamma(x)$. This is true for the sector of Figure 4 and hence for the union of all sectors starting from points of the set E. Of course, for points not belonging to the union of those sectors, we don't know anything. Let's look at the picture. What is the union of all these sectors over the points of E? It's very simple to visualize. Over each one of the component intervals of the complement you draw a "sawtooth", and the union Ω of the sectors is just everything that lies above those teeth (Figure 5).

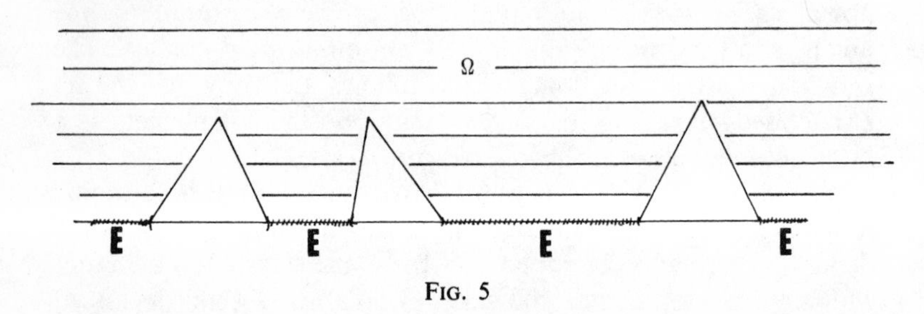

FIG. 5

That is, $\Omega = \cup_{x \epsilon E} \Gamma(x)$ is an open set on which $u(z)$ is a bounded, harmonic function. Unfortunately the boundary of Ω isn't very nice. It consists of an infinite number of teeth, together with straight lines. But at least it's Lipschitz so there is some hope of doing analysis on this domain. In one dimension the classical idea was to use conformal mapping. The domain Ω is after all simply connected; it is conformally equivalent to the upper half-plane. Now we have a function harmonic and dominated by one on the half-plane—a nice function on a nice domain. Therefore the behavior is very nice. There is still some rottenness; the conformal mapping is not as good as you'd like it to be, but at least you can hope to understand the behavior. This is the basic principle. You study potential theory on the sawtooth domain Ω.

Now there is a probabilistic model of the situation of Figure 5, which makes not only this particular theorem of Calderón, but ultimately the whole theory of H^p spaces, much clearer. I'll do it in one variable, but it can all be done just as well in n dimensions —whenever I say "interval" you should think "cube".

On the one hand there is the upper half-plane in the real world $R^2_+ = \{(x, y): y > 0\}$ on which we study harmonic functions. On the other hand, there is a model of the upper half-plane which I'll tell you about, on which we're going to study martingales.

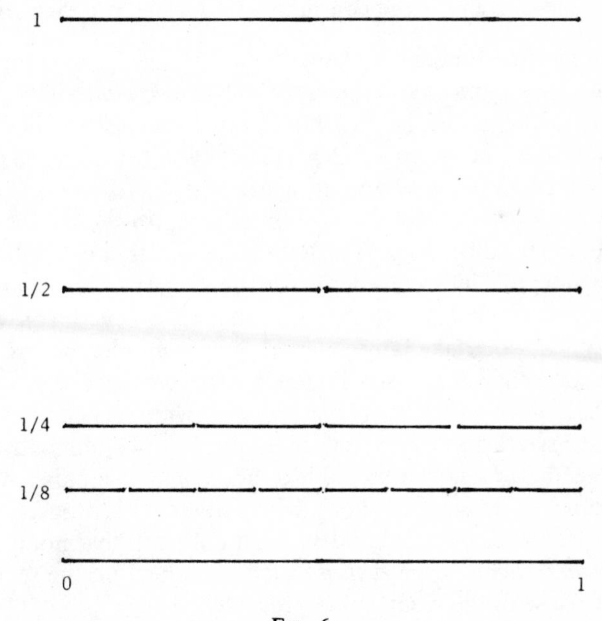

FIG. 6

What is a martingale? We'll see.

Given a function f on the line, you can form its Poisson integral. We won't care exactly what it is, but you find the Poisson integral $u_t(x)$ by convolving f with a certain peak with thickness roughly t.

Roughly speaking then, you start with some very complicated function f with peaks, singularities, and intricate oscillations, and to get $u_t(x)$, you smooth it out just enough so that it looks basically constant on little intervals of length t.

Now let's turn our attention to Figure 6, and find a martingale. We're going to start with a function on the interval $[0, 1]$. The countable family of line segments in Figure (c) is our discrete "upper half-plane". (At height $1/2^k$, the unit interval is cut into 2^k equal parts.) Now, given the function f, we're going to define its "Poisson integral" $u_k(x)$. On each interval, I'm simply going to average f, and $u_k(x)$ will be the averages at height 2^{-k}. For example, on the fourth and last segment of the third line of Figure 6, the value of $u_2(x)$ is the constant $4\int_{3/4}^{1} f(t)dt$. For each fixed k, if I compare the function $u_k(x)$ with the Poisson integral $u_{(2^{-k})}(x)$, there is a rough analogy between them. Because after all, the $u_k(x)$ also mimic the global behavior of f, but locally all oscillation is obliterated—a lot more crudely than by $u_t(x)$. The sequence $\{u_k(x)\}$ is an example of a martingale.

Let's see what Calderón's theorem is going to tell us about the discrete upper half-plane. As a slight generalization of the discrete Poisson integral, let's consider a function on the "half-plane" of Figure 6 which shares with Poisson integrals a key "mean value" property: If you take any interval (for example the left-most interval of height $1/2$—see Figure 6) and compare the number which you have assigned to that interval with the numbers that you have assigned to its two halves (to the first two intervals which are at height $1/4$ in our example), then the first number will be equal to the average of the other two numbers. This property holds for all "Poisson integrals" and so we'll define a "harmonic function" or *martingale* to be such a function defined on the segments of Figure 6 with the mean value property.

Now what is the analogue of Calderón's theorem? Given a point x in $[0, 1]$, the "non-tangential region" is the union of all the dyadic segments that lie directly above x, since the function $u(x, t)$ is constant on each dyadic segment of Figure 6. You would like to compare the statement that the $u_k(x)$ are bounded with the statement that they have a limit. If the $u_k(x)$ are bounded everywhere, then they have a limit almost everywhere. That's just the simple fact (proved by some elementary weak convergence argument) that every bounded martingale arises as the "Poisson integral" (in this crazy sense) of a bounded function f. The $u_k(x)$ are just the averages of the function, and then the Lebesgue theorem on the differentiation of the integral tells you that almost everywhere the $u_k(x)$ converge to $f(x)$. So the global theorem is trivial. Basically the same thing is true for the harmonic functions. Now let's suppose we're back in the general case—we don't have an f, but $\{u_k(x)\}$ stays bounded for x in a set $E \subset [0, 1]$ of positive measure. The analogue of Calderón's theorem says that for such a martingale $\lim_{k\to\infty} u_k(x)$ exists almost everywhere in E. Should one believe this? There's an analogue of the sawtooth picture here. In fact, it's nothing but the familiar Calderón-Zygmund decomposition which Hunt explains in his article. The complement of E is, let's say, the union of dyadic component intervals I_1, $I_2, \ldots$ (Figure 7).

The I_k are the largest dyadic intervals contained in the complement. For each point not belonging to I_k there is a sort of "sector" consisting of all segments lying above that point, where the martingale is bounded. Again, let's assume the bound α is uniform —that's not very important. What is the sawtooth region? Well, this time triangles have become squares, and the "sawtooth region" consists of all segments lying above dyadic intervals not entirely contained in the complement of E. And so the sawtooth region depicted in Figure 7 is obviously analogous to the one depicted in Figure 5. In the martingale case everything becomes simpler. I'd just like to point out that once you draw the sawtooth picture, it's rather trivial to see that everything is all right.

Perhaps I should point out that all this is really just a simple

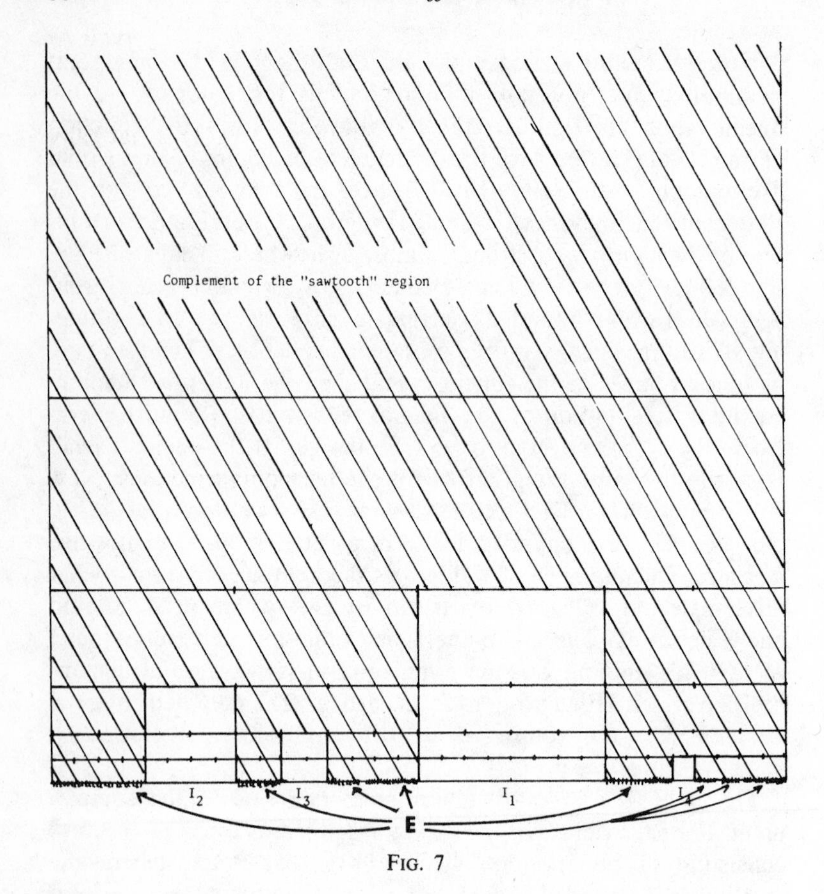

FIG. 7

exercise in probability theory. Consider the following game of chance. Let's say, in the simplest case, you have the "Poisson integral" $u_k(x)$ of a function f defined on the unit interval. Your fortune at time 0 is simply the average value of the function, $u_0(x)$. You may decide that you're happy with that and that you'll keep it; but if you decide to gamble, then you pass down to height $\frac{1}{2}$ (time 1), you flip a coin to determine which interval you're in, and then your fortune is the new average, i.e., the average over an interval (corresponding to one of the two segments) of length $\frac{1}{2}$. Again, you may be happy or you may decide to continue; if you continue, then you pick at random one of the two subintervals (of length $\frac{1}{4}$) and you continue gambling. At each stage, your fortune is simply the average over an appropriate dyadic interval. And the mean value property of the martingale says exactly that that game is fair. Note that if you decide to gamble at time k then your fortune changes by $(u_{k+1}(x) - u_k(x))$. In other words, the size of your bet is $|u_{k+1}(x) - u_k(x)|$.

You can stop gambling whenever you please. For instance, you might pick some large number α and quit as soon as the absolute value of the u_k exceeds α—that means that you've entered into one of the squares over an I_k (see Figure 7). That's a natural stopping time—it's something like "quit while you're (far) ahead (or far behind)". We'll keep that in mind.

Let me sketch the proof of the martingale version of Calderón's theorem. We have a martingale $\{u_k(x)\}$, a set E and a large $\alpha > 0$ with the property $\sup_k|u_k(x)| \leqslant \alpha$ for $x \in E$. To avoid technicalities, I'll suppose that the individual bets remain bounded. With more work, this extra hypothesis can be removed (see [5], p.320). By definition, u is a function on the discrete upper half-plane, constant on the top line of each of the white squares sitting on intervals $\{I_j\}$ in Figure 7. Let me now change u to a new martingale v by changing u on the white squares. Specifically, throughout the white square sitting on I_j I'll set v equal to the value of u at the top of the white square. At points of the discrete upper half-plane not contained in white squares, I'll set $v = u$. A moment's thought shows that v is a martingale, i.e., the mean-value property holds.

In terms of games of chance, v arises from u simply by halting all bets as soon as you enter a white square, i.e., as soon as the absolute value of your fortune exceeds α.

Calderón's theorem now follows from comparison of the two martingales u and v. On the one hand, $v_k(x) = u_k(x)$ for all k, if $x \in E$. In other words, if your fortune never exceeds α, then the betting never stops, and $v_k(x) \equiv u_k(x)$. So to prove that $\{u_k(x)\}$ converges for $x \in E$, it is certainly enough to show that $\{v_k(x)\}$ converges a.e. On the other hand, v is a bounded martingale, since you simply stop gambling as soon as your fortune grows too big.

Consequently the a.e. convergence of $\{v_k(x)\}$ follows immediately from the easy global convergence result, and Calderón's theorem is proved.

Suppose that you wanted to know a condition for the existence of the limit of the martingale in terms of the size of the bets that you make. It's pretty clear that if you play at matching pennies and you bet a dollar every time, your fortune will not remain bounded. On the other hand, if you bet some ever smaller amount of money, let's say decreasing in a geometric series, you have no chance to diverge. This is connected to the following classical problem. Everyone knows that the harmonic series $\Sigma\, 1/n$ diverges, but if you attach alternating plus and minus signs, the resultant series

$$\sum_{n=1}^{\infty} \frac{(-1)^{n+1}}{n}$$

converges (to ln 2). The question is, suppose that you pick the sequence $\{\epsilon_n\}$ at random by flipping a coin (if heads, let $\epsilon_n = +1$; if tails, let $\epsilon_n = -1$); what is the probability that $\Sigma_{n=1}^{\infty} \epsilon_n / n$ converges? (Incidentally the martingale for this game is

$$\begin{array}{lcccc}
u_0: & \multicolumn{4}{c}{0} \\
\hline
u_1: & \multicolumn{2}{c}{1} & \multicolumn{2}{c}{-1} \\
\hline
u_2: & 1+\frac{1}{2} & 1-\frac{1}{2} & -1+\frac{1}{2} & -1-\frac{1}{2} \\
\hline
 & \vdots & \vdots & \vdots &
\end{array}$$

so that $u_3(x) = 1 - (1/2) + (1/3)$ if $3/8 < x < 1/2$, and so forth.) The classical answer is that not only does this converge with probability 1, but it isn't even close. The sharp result has to do with things like $\Sigma_{n=1}^{\infty} \epsilon_n / n^{\frac{1}{2}+\delta}$, which converges with probability one (for $\delta > 0$). This has to do with the size of the bets. This is a simple example of a strategy that you could use at our "martingale game". You could, of course, use much fancier strategies—you could decide to bet one cent until, let's say, you either won or lost \$1000. Then if you've won the \$1000 you quit, but if you've lost the \$1000, then you become very angry and bet 2^n dollars at time n. Here our martingale starts

u_0:		0		
u_1:	.01		$-.01$	
u_2:	.02	0	0	$-.02$

and proceeds in orderly fashion until we read the line of height $2^{-100,000}$. For $k \geqslant 100,000$, all the $u_k(x) = 1000$, on the interval $0 < x < 2^{-100,000}$. However, if x is very close to 1 (say $1 - x < 2^{10^9}$), then (say) $u_{-100,007}(x) = -1000 - 1 - 2 - 4 - \cdots - 2^7$, and so on. You can do virtually anything.

Now it's pretty clear that if you adopt a strategy with big bets, your fortune fluctuates wildly. If you adopt a strategy with small bets, your fortunes are much more likely to converge. But for the existence of a limit, it isn't very important how lucky you are; because except for miracles—events of probability zero—you expect that things will sort of average out and the only thing that should be really important is the size of your bet. That means that E, the set where the maximal function is finite, $(E = \{x | sup_k |u_k(x)| < \infty\})$, should be the same as the set where something else is finite—a thing which has to do only with the size of the bets. What is that thing? It is the square function, the sum of the squares of the bets. In other words,

$$\left\{x \middle| \sup_k |u_k(x)| < \infty\right\}$$

$$= \left\{x \middle| \Sigma_{k=1}^{\infty}(u_k(x) - u_{k-1}(x))^2 < \infty\right\} \text{ except for a set of measure zero.}$$

The quantity $\left(\Sigma_{k=1}^{\infty}(u_k - u_{k-1})^2(x)\right)^{1/2}$ is called the martingale square function, or S-function. The analogue of this in the case of the upper half-plane R^2_+ is something that has been written down before. It's the classical Lusin area integral. After all, S has to do with the rate at which the harmonic function changes. If it's changing very rapidly at the k-th step—that means that you have a large bet ($= |u_k(x) - u_{k-1}(x)|$) at the k-th step—in other words that u has a large gradient. This means that the analogue of S to be expected is

$$S(u)(x) = \left(\int\int_{\Gamma(x)} |\nabla u|^2 d\sigma \right)^{1/2}.$$

(There are some fudge factors to be put in the integrand in n dimensions, but nothing really changes.) The point that I'd like to make is that based on probabilistic ideas, it's entirely reasonable that $S(u)$ should be equivalent to the maximal function $u^*(x) = \sup_{\Gamma(x)}|u|$. So $S(u)$ and u^* are equivalent in some very strong sense. We've seen, for instance, that (at least it seems plausible here) except for a set of measure zero, the set where $S(u)$ is finite should be the same as the set where u^* is finite. By the way, if you want to prove at least the dyadic probabilistic version of this fact, it's very easy from our martingale picture. Similarly, by somewhat fancier techniques, one can prove the same thing using the sawtooth picture (Figure 5). But what I'd like to concentrate on is the L^p norms of $S(u)$ and u^* are the same, $\|u^*\|_p \sim \|S(u)\|_p, 0 < p < \infty$. This is the theorem of Burkholder, Gundy, Silverstein.

Let me take a little bit more time to mention a very charming point—the way that the original proof went. Nowadays the proof is based on the upper half-plane—martingale analogy, but originally it was done actually by an almost literal translation. Martingales are fair games in general. (There is an official, rigorous definition.) In fact, there is not just an analogy, but even a martingale associated canonically to a harmonic function. I'd just like to say a little about it, and then drop it. Suppose you have a harmonic function u in some domain. Take a particle, put it at some interior point—an arbitrary starting point—and let it undergo Brownian motion. It moves around at random—strange things happen. Call its position at time t, X_t. (X_t is a random variable, an undetermined point.) You can consider the value of u at X_t—that's a martingale. Why? Let's say that the particle is now at position X_t, so that your fortune is $u(X_t)$. Suppose you decide to go on to some slightly later time $t + \Delta t$. You look at $u(X_{t+\Delta t})$—then the average value of your fortune later (at time $t + \Delta t$) is equal to your fortune now (at time t), which means that the game is fair—on the average you neither gain nor lose. That's very simple, for after all, how has the particle drifted from X_t? Well, there's some highly peaked Gaussian distribution centered at X_t. The main point is that the distribution has circular symmetry. So to average $u(X_{t+\Delta t})$ given X_t, is to take a weighted average of averages of u over circles centered at X_t. Therefore $\{u(X_t)\}$ is a martingale, by the mean value property of the harmonic function $u(x) = (1/2\pi)\int_0^{2\pi} u(x + \rho e^{i\theta})d\theta$ for any $\rho > 0$ so small that the circle $x + \rho e^{i\theta}$, $0 \leqslant \theta < 2\pi$, is contained in the region where u is harmonic [1]. So in this way you get a martingale. Burkholder, Gundy and Silverstein first proved $\|u^*\|_p \sim \|S(u)\|_p$ by looking carefully at the martingale $\{u(X_t)\}$ and proving analogues of the theorems for the martingales $\{u_k(x)\}$ discussed above. In any event, $\|u^*\|_p \sim \|S(u)\|_p$ is true. So what?

Well, the equivalence $\|u^*\|_p \sim \|S(u)\|_p$ already has some non-trivial information in it. Let's already look at the easy case where p is bigger than 1. In that case $\|u^*\|_p$ is finite as soon as the boundary values of u belong to L^p. This follows just from the ordinary maximal theorem for L^p. So we know that $\|u^*\|_p < \infty$ if and only if $u =$ P.I. (f), with f in L^p. On the other hand, our equivalence says $S(u) \in L^p$. This already contains the L^p boundedness of the Hilbert transform—that's the only point I'm making here. Because, after all, compare u with the conjugate function $v =$ P.I.(g). We'd like to know whether v is in L^p. Well, this equivalence theorem says that instead of comparing u with v (which is difficult), it's enough to compare the S functions, $S(u)$ with $S(v)$. If you stare at the definition, the Cauchy-Riemann equations tell you that the absolute value of the gradients are the same:

$$\frac{\partial u}{\partial x} = \frac{\partial v}{\partial y}, \quad \frac{\partial u}{\partial y} = -\frac{\partial v}{\partial x}$$

so that

$$\left|\left(\frac{\partial u}{\partial x}, \frac{\partial u}{\partial y}\right)\right| = \left|\left(\frac{\partial v}{\partial x}, \frac{\partial v}{\partial y}\right)\right|,$$

and so the S functions are identical, point for point. So it's trivial that $\|S(u)\|_p = \|S(v)\|_p$, and therefore by the equivalence, u and v have comparable L^p norms on the boundary. So in this trivial case you get interesting information. I just used that to show you that there is some information contained in these ideas. But there is much more than that. Let's come back to the Cauchy–Riemann systems. We have a vector of harmonic functions $F = (u_0, u_1, \ldots, u_n)$—if you like, perhaps just an analytic function in the half-plane, $F = u + iv$, for the moment. If $F = u + iv$ is in H^p, then we've seen that the maximal theorem tells you that F^* is in L^p. Let's drop all the information that that gives us about v and let's just look at u. So of course $u^* \in L^p$. The theorem of Burkholder, Gundy, Silverstein says that all this can be reversed, that u^* in L^p implies F is in H^p is true too (so that $F \in H^p \Rightarrow F^* \in L^p \Rightarrow u^* \in L^p \Rightarrow F \in H^p$). For after all, if $u^* \in L^p$, by the same trick that enabled us to bound the Hilbert transform, it's clear that v^* must belong to L^p also, and in particular then you have boundary values in L^p, and so F must belong to H^p. So what? Well, now you don't have to think of an element of H^p as being either an analytic function or a vector of functions satisfying a system of Cauchy-Riemann equations. It's actually enough just to look at ordinary harmonic functions. Thus here is a natural condition which is necessary and sufficient for u to be the real part (in the case of the half-plane); or to be the u_0 $\left(\textit{in } R^{n+1}_+\right)$ for a Cauchy-Riemann system in the Stein-Weiss H^p. So now we have a new definition: Given a harmonic function u, $u(x, t)$ *belongs to*

H^p if and only if the maximal function $u^*(x)$ belongs to L^p. This is satisfying if for no other reason than that the definition doesn't depend on p. (Recall that the system of Cauchy-Riemann equations changes as p drops toward 0.) Another advantage is that you can draw an analogy between this definition and something dyadic, so that one can play with probabilistic models and see by analogy what's going on in H^p. On the other hand, there seem to be no very convincing analogues of the Reisz transforms. Now I'd like to tell you an even newer definition of H^p.

If you think of a harmonic function u as the Poisson integral of some f on the boundary, $u =$ P.I. (f), perhaps u is not very nice and then f is merely a distribution, but at least u arises as the convolution of something with the Poisson kernel. Now it turns out by some results of Stein and others (editor's reference: [7]) that there is nothing very special about the Poisson kernel. Actually you could take any old approximate identity $\{\phi_t\}$ and form the "Poisson integral", which is simply $u(x, t) = (\phi_t * f)(x)$. You could then take the maximal function by setting $u^*(x) = \sup_{(y,t)\epsilon\Gamma(x)} u(y, t)$. To require u^* to belong to L^p for any approximate identity which is at all reasonable, is in fact equivalent to our old new definition. In other words, our newest definition is that *u is in* H^p if and only if u arises as the convolution of a distribution with some "reasonable" approximate identity in such a way that the resultant maximal function is bounded. (Say that an approximate identity $\{\phi_t(x)\}$ on R^n is "reasonable" if $\phi_t(x) = t^{-n}\phi_1(x/t)$ with $\phi_1(x)$ smooth and "small" at infinity. Thus, the "thickness" of ϕ_t is t.)

Our latest definition tells us something new. We started out with analytic functions or vectors of harmonic functions satisfying differential equations. Some structure was removed initially by our first new definition. By that, we saw that H^p is really just a set of harmonic functions. But now, since our latest definition doesn't depend on the particular choice of approximate identity, actually H^p is just a creature of general nonsense, of elementary real variables. It's simply the space of all distributions on the boundary for which some maximal function belongs to L^p.

Once we've gotten to this stage, then it's possible to go back and reprove some of the more classical theorems and everything is much simpler. The reason is as follows. Recall that the usual classical theorems of Fourier analysis, the results of Littlewood-Paley, the boundedness of the Hilbert transform, of singular integrals in R^n, and so forth—these results are valid for p strictly between 1 and $\infty (1 < p < \infty)$. The big trouble with $p = 1$ and even more so with $p < 1$ is that you don't have the maximal theorem. (There is a weak type (1, 1) inequality, but that doesn't help. Starting with f in some strong class (for example L^1), you only get that f^* belongs to a weak class,

$$\text{weak } L^1 = \{ g | \text{measure}\{ x | \, |g(x)| > \alpha \} < c(g)/\alpha \text{ for all } \alpha > 0 \}.$$

Now what I'm saying in effect is, "All right, let's make ourselves a Hardy-Littlewood theorem by fiat, by saying that H^p consists of distributions whose maximal functions are in L^p". So we've defined the Hardy-Littlewood theorem to be true. Now everything else follows from it—and, in fact, rather simply. For you can now carry out the Calderón-Zygmund decomposition for functions in H^p. That is, you can take the picture of Figure 7 and cut up f into a good piece and a bad piece. (Admittedly, you have to be a little more careful by averaging f with smooth partitions of unity instead of disjoint intervals. But that's just a technical point.)

Let me just look at one classical example. Let's say we're on the line. Consider a blip f which has thickness δ and height δ^{-1} (Formally, f can be any reasonably smooth function satisfying $f(x) \geqslant 0$, support $(f) \subset (-\delta, \delta)$, $\sup|f| \leqslant \delta^{-1}$, and $\int f = 1$; for example, one might take

$$f(x) = \frac{1}{\delta} \exp\left[c - \frac{c\delta^4}{(x^2 - \delta^2)^2} \right], x \in (-\delta, \delta),$$

where $c = .7674537635\ldots$ is determined by the $\int f = 1$ condition.) I'm going to be looking at H^1 for simplicity. This function is in L^1, but it's not in H^1. Why? Let's chase it through all the different stages of the definition. We're thinking of this as the boundary value of the real part of an analytic function. The analogous imaginary part is the Hilbert transform of f. The Hilbert transform has a $1/x$ singularity, so $\tilde{f}$ looks basically like $1/x$ until you get to within a distance of δ or 2δ or so of the origin. (Of course $\tilde{f}$ becomes smoother than $1/x$ near 0.)

Since $\tilde{f}$ behaves like $1/x$ away from the origin, the integral of its absolute value doesn't converge. So, visibly, it isn't in L^1—its $1/x$ singularity introduces a nasty logarithm, which ruins everything. On the other hand, suppose you construct a different sort of function, call it g, by taking a blip of height $\frac{1}{2}\delta^{-1}$ and thickness δ on $(0, 2\delta)$ and extending it to be an odd function. Explicitly, one may set $g(x) = f(x - \delta)/2$ on $(0, 2\delta)$. The function g has some cancellation in it, but is otherwise the same as f. This odd function has integral zero. So of course the Hilbert transform is much smaller than $\tilde{f}$. For example, consider the value of $\tilde{g}$ at x_0 (where x_0 is large). To compute $\tilde{g}(x_0)$ you multiply pointwise these two functions, $g(t)$ and $1/(t - x_0)$

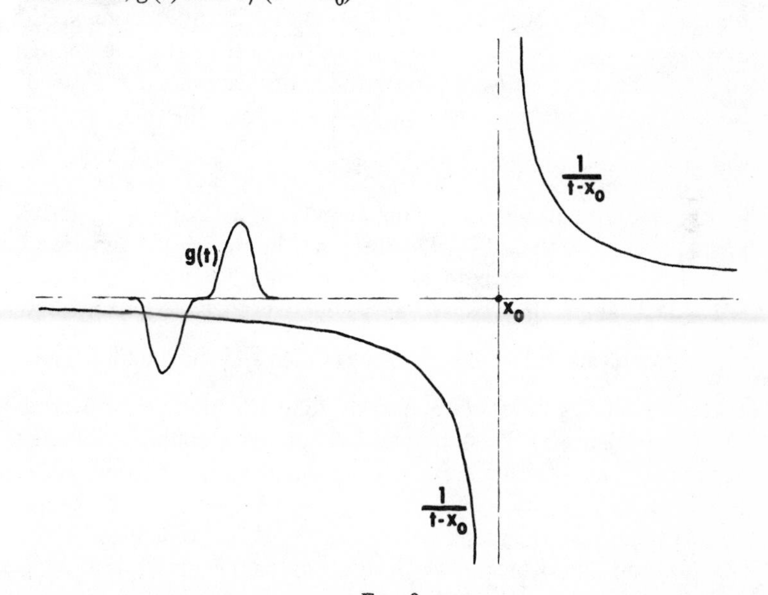

FIG. 8

You then integrate over $-\infty < t < \infty$. It's true that the absolute area under g is 1, but on the hand, the part of g below the x axis essentially cancels out the part above the axis. Since $1/(t - x_0)$ is almost constantly equal to $-1/x_0$ throughout $(-2\delta, 2\delta)$ and so the value of $\tilde{g}(x_0)$ is much smaller than $1/x_0$, there is a much faster decrease, more like $1/x^2$, or even a higher negative power of x.

So $\tilde{g}$ is in L^1. In fact, g is a typical function in H^1—something with cancellation. (Let me pause and remind you again that I'm now using the term "H^1 function" for what used to be the real part of the boundary value of an analytic function $F \in H^1$. So do not be alarmed when g has compact support despite theorem (c) above.) Let's look at this in terms of the next definition. Let's look at the Poisson integrals of f and g and compare them. In fact, already let me go to the general nonsense definition. Given any large number x_0, you integrate f or g against some approximate identity ϕ_t of "thickness" t and peaking at x_0—you're free to vary the t—the thickness of the peak (since we're interested in the maximal function $\sup_t \phi(x, t)$). It's clear that in order to make the average as big as possible you want to make the t just big enough so that the support of f falls under the main mass of the peak of ϕ_t. So the right t to take is just $|x_0|$. If you do that, then you see that the maximal function $u^*(x_0) = \sup_{t>0}\left|\int_{-\infty}^{\infty} \phi_t(x - x_0) f(x) dx\right|$ looks something like $1/x_0$. Again we get a $1/x$ kind of singularity, just as was the case with $\tilde{f}$. On the other hand, for the function g you get cancellation as before. Again the right kind of peak to take is $t \sim |x_0|$, again the positive and negative parts of g contribute cancelling terms to $\int_{-\infty}^{\infty} \phi_t(x - x_0) g(x)\, dx$ and the maximal function is decreasing much more quickly than $1/x$ and hence belongs to L^1 (see Figure 8). To summarize, a function with antisymmetric peaks (such as g) is the sort of thing that we expect to belong to H^1.

This same thing happens in R^n with the Riesz transforms. For a function on R^n which, like f, has no cancellation, the Riesz transforms have the wrong kind of singularity (namely the $R_j f$ are too big at infinity to belong to $L^1(R^n)$), but if you take a function with finite support and average zero (like g), then there will be enough cancellation to make the Riesz transforms belong to L^1. Let me restate this whole thing again for the n-th time. Let's say that among all L^1 functions, we'd like to single out a good class to which singular integrals can be applied. Most L^1 functions are very bad. There is no reason to assume that a random L^1 function would have much cancellation. And then you see that things are bad. You apply virtually any sort of singular integral operator to a function like f and it doesn't work. On the other hand, there is H^1, the class of good L^1 functions. To this you can apply lots of operators and the images of these H^1 functions (for example, the Hilbert transform of g) are in L^1. The philosophy is that there is a dichotomy. You hope that any L^1 function is either bad, which means essentially whatever you do to it you wind up outside of L^1; or it's good, in which case more or less whatever you do to it, you stay in L^1. We've seen that for the Riesz transforms, that works. That is, the Riesz transforms form a test. Given a function f in $L^1(R^n)$, apply the Riesz transforms and check to see whether the n Riesz transforms belong to L^1. If they don't, then already just applying a particularly simple system of singular integrals, you already leave the class L^1. But if these test operators send our function back into L^1, that means that all singular integrals will work—the S function will work—the maximal function will work —and so forth and so on.

So here is a natural conjecture. I don't know how to prove it, but it expresses this dichotomy. I don't claim it's original. I'm sure lots of people have thought about it. I'm perhaps the first person foolish enough to stick my neck out and give it in public.

CONJECTURE: Let $f \in L^1$ and let $K_1, K_2, \ldots, K_N$ be a finite collection of singular integral kernels $K_i(x) = \dfrac{\Omega_i\left(\frac{x}{|x|}\right)}{|x|^n}$, where Ω_i has mean value zero and is very nice. (C^∞ if you like.) Suppose that the K_j's are somehow non-degenerate—for example we could assume that the $\hat{K}_j(\xi)$ are elliptic (have no common zeros on the set $\xi \neq 0$). Finally suppose that $K_j * f \in L^1$ for every j. Then f is in H^1.†

If this conjecture is true, it would mean that indeed an L^1 function is either quite nice or quite rotten. Then it would be clear that H^1 is a very natural class to single out for study since it would be exactly that class of functions for which you can do anything. It's rather obvious that if the K_j's are small perturbations of the Riesz transforms, that is if the Ω_j's as functions in C^∞ on the unit sphere S^{n-1} are small perturbations of the corresponding Ω_j's for the Riesz transforms; then there is no problem in proving the conjecture by a perturbation argument. But other than that, I don't know whether the conjecture is true. Not only don't I know how to prove it, I'm not entirely sure how confident one should be of it.

Finally, I'd like to say a little about functions of bounded mean oscillation. This is a technique by which you can prove theorems about H^1. This particular technique tells you nothing about H^p

---

†The author made this conjecture in his talk in June 1974. Less than 6 months later, I received a counterexample from J. Garcia-Cuerva, who is studying at Washington University of St. Louis. He sets

$$N = 2, \quad K_1(x_1, x_2) = \frac{x_1^2 - x_2^2}{|x|^4}, \quad K_2(x_1, x_2) = \frac{-2x_1x_2}{|x|^4},$$

and constructs a purely radial function $f(x)$ such that $f \in L^1(R^2)$, $K_i * f \in L^1(R^2)$, but $f \notin H^1(R^2)$. Let $z = x_1 + ix_2$ and note that $K_1(z) + iK_2(z)$ is simply $1/z^2 = e^{-2i\theta}/r^2$. His function $f(r)$ is an infinite linear combination of functions of the form

$$n\chi_{[n, n+1/n]} - \frac{1}{n}\chi_{[n+1/n, 2n+1/n]}$$

where $\chi_{[a, b]}$ denotes the characteristic function of the interval $a \leqslant r \leqslant b$. Garcia-Cuerva's computations are somewhat involved, but deftly accomplished.

Since the set $\{(\cos\theta)/r^2, (\sin\theta)/r^2\}$ does test (being the Riesz transforms) and, as we have seen, the set $\{(\cos 2\theta)/r^2, (\sin 2\theta)/r^2\}$ does not, something very subtle is going on here. Fefferman has suggested to me that perhaps the closure in the sup norm of the algebra generated by $\{\Omega_1, \ldots, \Omega_N\}$ must be the set of all continuous functions on the torus. —*Editor.*

for $p < 1$, but sometimes you can guess that a certain theorem true for H^1 might also be true for $H^p, p < 1$, and then try to prove it by other methods. On the one hand, it's possible to compute the dual of H^1. On the other hand, the dual is something so simple that given an operator which you would like to be bounded on this dual, it seems never to be very hard to check whether it is.

The dual class of functions is just this. Let's work on the line. (As always the same thing can be done on R^n by saying "cube" instead of "interval".) Given an interval I and a locally integrable function f, you consider the average $f_I = (1/|I|)\int_I f(x)\,dx$. Then you subtract from f its average f_I, so that the difference $f(x) - f_I$ has mean value 0 over I

$$\left(\int_I (f(x) - f_I)\,dx = \frac{1}{|I|}\int_I f(x)\,dx|I| - f_I\int_I dx = 0\right).$$

Now form the expression $(1/|I|)\int_I |f(x) - f_I|dx$. This is some measure of how much the function oscillates over the interval. If

$$\sup_I\left(\frac{1}{|I|}\int_I |(f(x) - f_I)|dx\right)$$

is finite, where you consider all intervals, no matter where they're placed, or how big or how small they are, then f is said to be of bounded mean oscillation, $f \in$ B.M.O. I'd like to say just a little bit about it. Is belonging to B.M.O a strong condition or a weak condition? It's very strong. Bounded functions obviously belong to B.M.O. (If a function is bounded by 1, then $|f(x) - f_I|$ is bounded by 2, the average is bounded by 2—everything's as nice as could possibly be.) So the interesting question is, which unbounded functions belong to B.M.O.? Let's try some examples. The function $|x|^{-\delta}$ doesn't work no matter how small $\delta > 0$ is. The reason is that although $|x|^{-\delta}$ could be normally considered very nice, nevertheless here you're measuring its behavior on arbitrary intervals, and on the particular interval $[-.01, +.01]$ the function doesn't look very good. This indicates that the property of belonging to B.M.O. is something quite strong. The typical example of a function of bounded mean oscillation is the log, let's say $\log 1/|x|$. (This is the right sort of singularity. You can check that it works.) So functions of bounded mean oscillation have rather weak singularities. It's a theorem of John and Nirenberg that in fact, at least in terms of the distribution function—the measure of the set where the function is big—$\log 1/|x|$ is the worst possible example. But there is a little more to it than that. Namely, if you look at the function $s(x) = \text{sgn}(x) \log 1/|x|$ which, in terms of size, is exactly the same as $\log 1/|x|$—then visibly $s(x)$ is not of bounded mean oscillation. For, if you take a very tiny interval about the origin, then it's clear that there is no constant S_I that comes close to $s(x)$ throughout the interval. In summary, a function of bounded mean oscillation has to have at worst a singularity on the order of $\log 1/|x|$ and the peaks have to be symmetric. Now the theorem is that this Banach space of functions (B.M.O.) is the dual of H^1 (editor's reference: [6]). I'd just like to point out that although the different proofs involve various tricks; you can guess that it ought to be true if you look at the graphs of $\log 1/|x|$ and $s(x)$. After all, what kind of functions can you integrate against H^1? Remember that $g(x)$ is in H^1, while $f(x)$, which doesn't have its cancelling properties (recall that g was defined on p. 69), isn't in H^1. Of course H^1 functions are L^1, so certainly anything L^∞ can be integrated against $g(x)$. What about an unbounded function? Well, things can't be too bad, but perhaps a mild singularity could be allowed. But you have to be on guard, because if you have, let's say, an antisymmetric singularity (such as $s(x)$), you're in trouble. You multiply together $g(x)$ and $s(x)$ and the integral is infinite—you lose the advantage of the cancellation. In order to preserve the advantage given to you by the cancellation, you have to see to it that if the candidate for $(H^1)^*$ has a logarithmic peak right at the origin where $g(x)$ changes sign, then that peak must be symmetrical—just like $\log 1/|x|$.

In closing, I'd just like to mention that rather recently—in the last month or so, Carleson has discovered a very interesting counterexample which proves that much of this stuff collapses in the polydisc. And so, there's some question of what to do in the polydisc. Perhaps our points of view should be changed with the polydisc in mind so that something will go over. In summary, then, it's not at all clear whether we really have the right approach to H^p.

REFERENCES

1. Burkholder, D. L., R. F. Gundy, and M. L. Silverstein, "A maximal characterization of the Class H^p," *Trans. Amer. Math. Soc.*, **157** (1971), 137–153.
2. Calderón, A., "Commutators of singular integral operators," *Proc. Nat. Acad. Sci.*, **53** (1965), 1092–1099.
3. Calderón, A., and A. Zygmund, "On higher gradients of harmonic functions," *Studia Math.*, **26** (1964), 211–226.
4. Coifman, R. R., and Y. Meyer, "On commutators of singular integrals and bilinear singular integrals," *Trans. Amer. Math. Soc.*, **212** (1975), 315–331.
5. Doob, J. L., *Stochastic Processes*, Wiley, New York, 1953.
6. Fefferman, C. L., "Characterizations of bounded mean oscillation," *Bull. Amer. Math. Soc.*, **77** (1971), 587–8.
7. Fefferman, C. L., and E. M. Stein, "H^p spaces of several variables," *Acta Math.*, **129** (1972), 137–193.
8. Stein, E. M., and G. Weiss, "On the theory of harmonic functions of several variables, I. The theory of H^p spaces," *Acta Math.*, **103** (1960), 25–62.
9. Zygmund, A., *Trigonometric Series*, (2nd edition), 2 vols., Cambridge, England, 1959.

MULTIPLE TRIGONOMETRIC SERIES

J. Marshall Ash

1. INTRODUCTION

The purpose of this article is to give the reader a feeling for the obstacles that arise when one tries to extend the far-reaching and well-developed theory of one dimensional trigonometric series to higher dimensions. I readily confess to slanting this talk towards areas I have worked in and make no claim of comprehensiveness. Furthermore, I will resist the temptation of stating a best known or "best possible" result whenever a less good result is easier to understand but still captures the spirit of the situation. (For example, the hypothesis of $f \in L^p$ might be used when a weaker hypothesis such as $f \in L(\log^+ L)^2$ would be sufficient.) In such cases, I will try to give enough references so that the interested reader can trace down the stronger version of the result.

I will mainly explicate the theory of double Fourier series, whose generality and difficulty is intermediate between the *terra firma* of one dimension and the *terra incognita* of three dimensions. At the present time, the passage from two to three dimensions seems far more substantial and non-trivial than that from one to two or than that from three to more.

Let $f(x, y)$ be a measurable complex valued function defined on the unit square $T^2 = [-\frac{1}{2}, \frac{1}{2}) \times [-\frac{1}{2}, \frac{1}{2})$.

If the Lebesgue integral $\iint_{T^2} |f(x, y)|^p \, dx \, dy$ (where $p \geqslant 1$) is finite, we say $f \in L^p$. In this case, f has a Fourier series given at each (x, y) of T^2 by

$$S[f](x, y) = \sum_{m, n} \hat{f}_{mn} e^{2\pi i(mx + ny)}, \tag{1.1}$$

where the Fourier coefficients $\hat{f}_{mn}$ are given by

$$\hat{f}_{mn} = \iint_{T^2} f(s, t) e^{-2\pi i(ms + nt)} \, ds \, dt, \tag{1.2}$$

and the summation is indexed by the m-n lattice plane—the set of all ordered pairs of integers. This Fourier series is a very special case of a double trigonometric series where the $\hat{f}_{mn}$ are replaced by arbitrarily chosen complex constants c_{mn}. As in the one dimensional case, the set of all double Fourier series is only a very, very, tiny subset of the set of all double trigonometric series.

The lattice plane has no natural ordering and many important trigonometric series are not absolutely convergent, so the first order of business is to determine an ordering.

We will take the terms corresponding to the four points (m, n), $(m, -n)$, $(-m, n)$, and $(-m, -n)$ together (just as one takes $\hat{f}_n e^{2\pi inx} + \hat{f}_{-n} e^{-2\pi inx}$ as a term in the one dimensional theory). Even this is less natural than in the one dimensional case as our later discussion of the failure of Plessner's theorem will indicate.

This grouping of the four terms reduces the problem of ordering the lattice plane to that of ordering the lattice quadrant. How shall we do this? The answer is far from clear. In fact, there very probably is no one answer.

To get a feel for the situation we will consider some numerical series that ought not to converge, but which do with respect to some of the usual orderings.

Let $s_{m, n} = \sum_{(i, j) \text{ southwest of } (m, n)} a_{ij}$ be the rectangular partial sums.

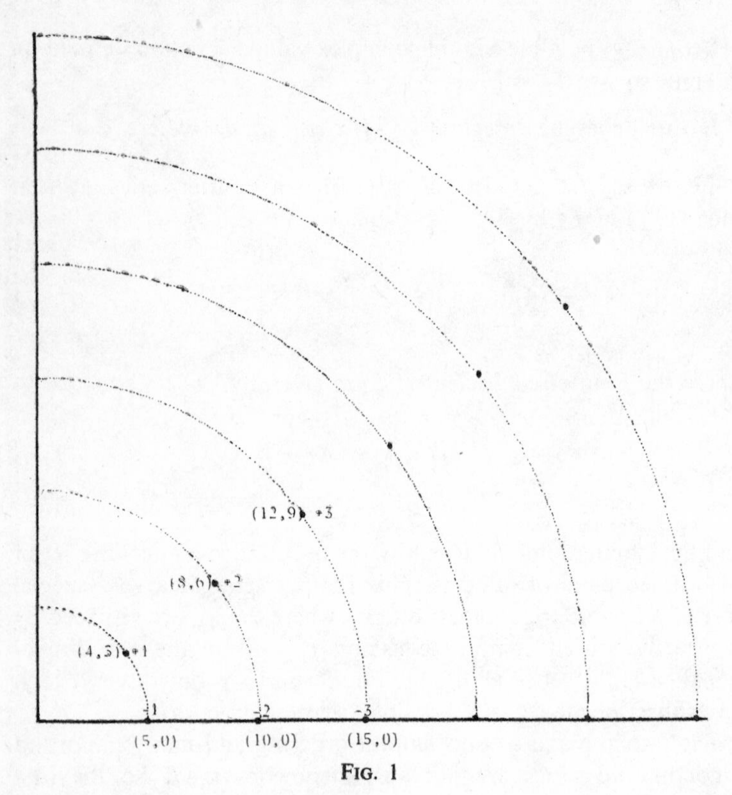

FIG. 1

For example,

$$s_{3,1} = a_{0,1} + a_{1,1} + a_{2,1} + a_{3,1} + a_{0,0} + a_{1,0} + a_{2,0} + a_{3,0}.$$

Example 1. Let $a_{5k,0} = -k$, $a_{4k,3k} = k$, $a_{mn} = 0$ otherwise (Figure 1).

A numerical double series is *square convergent* if the limit of s_{nn} exists as n tends to infinity. Here $s_{5k-1,5k-1} = k$ so $\lim_{n\to\infty} s_{nn}$ doesn't exist. This series is not square convergent. Let $s_r = \sum_{\sqrt{m^2+n^2} \leq r} a_{mn}$. Then a numerical series is *circularly convergent* if the limit of s_r exists as r tends to infinity. However, in example 1, for all r, $s_r = 0$, so $\lim_{r\to\infty} s_r = 0$, so the series is circularly convergent.

Example 2. Let $a_{kk} = -k$, $a_{k0} = k$, $a_{mn} = 0$ otherwise.

Here $s_{nn} = 0$ for all n so the series is square convergent. (Draw a picture.)

A third method of convergence is restricted rectangular convergence. Here we are interested in the "limit" of the not too eccentric rectangular partial sums—rectangles that are "sort of" close to squares. Since the set of all such rectangles is not linearly ordered the definition requires some fancy footwork. We say that the series *converges restrictedly rectangularly* to s if for every fixed $E > 1$, $\sup_{(m,n) \in W_E(N)} |s_{mn} - s| \to 0$ as $N \to \infty$, where $W_E(N)$ ($= \{(m, n) : m \geqslant N, n \geqslant N, Em \geqslant n \geqslant E^{-1}m\}$) are a sequence of wedges cut out of the lattice quadrant shown in Figure 2.

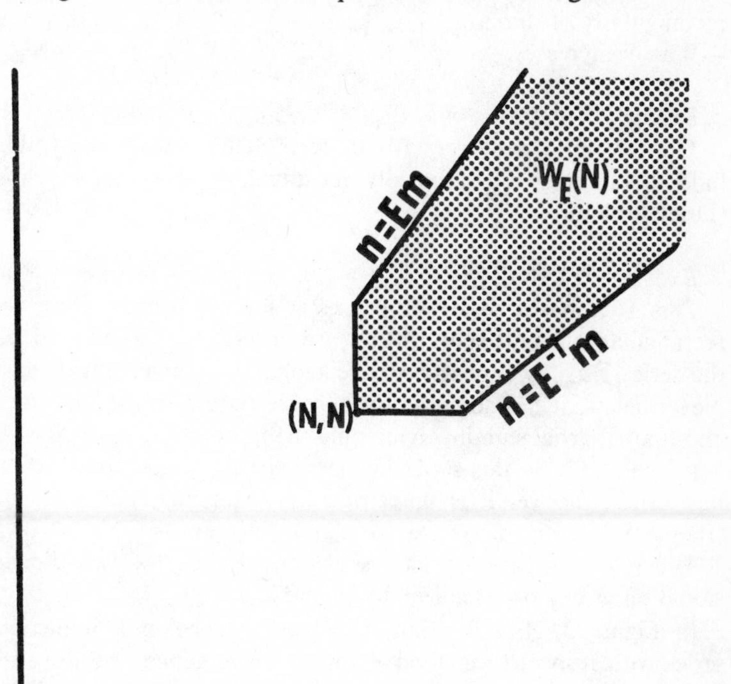

FIG. 2

The intuitive idea behind this picture is that if E is not too much bigger than 1, then a rectangle with southwest corner at (0, 0) and northeast corner in $W_E(N)$ is pretty nearly a square. Furthermore, the point of the N is that as N becomes very large you are looking at a "late" partial sum.

It is equivalent to demand that $s_{m_i, n_i} \to s$ for every sequence (m_i, n_i) tending to (∞, ∞) in such a way that the two ratios m_i/n_i and n_i/m_i remain bounded.

Observe that example 2 is not restrictedly rectangularly convergent since $s_{k, k-1} = k$.

The last method of convergence I will discuss is that of *unrestricted rectangular convergence*. We say $s_{mn} \to s$ unrestrictedly rectangularly if $\lim_{\min\{m, n\} \to \infty} s_{mn} = s$, i.e., if $\sup_{m, n \geqslant N} |s_{mn} - s| \to 0$ as N increases.

Example 3. Let $a_{k^2, 0} = k$, $a_{k^2, k} = -k$, $a_{mn} = 0$ otherwise.

This example converges (to 0) restrictedly rectangularly, but fails to converge unrestrictedly rectangularly since $s_{k^2, k-1} = k$. (Draw a picture.)

Example 4. Let $a_{k, 0} = k$, $a_{k, 1} = -k$, $a_{mn} = 0$ otherwise.

This is a nasty series—it causes a lot of trouble. Here the rectangular sums, as soon as they are at least $n \times 2$, are zero. So the series Σa_{mn} is unrestrictedly rectangularly convergent to zero. Nevertheless, it has some pretty horrible partial sums. Just think about any partial sum involving only the bottom line $s_{n0} = \Sigma_{i=0}^{n} a_{i0} = n(n+1)/2$. So this series has terribly bad partial sums; it has big terms, and yet it is unrestrictedly rectangularly convergent. (Incidentally, it is, of course, circularly divergent.)

The relations between modes of convergence for two dimensional series can be visualized by Figure 3.

In Figure 3, $A \to B$ means that convergence of a numerical series with respect to method A forces convergence of that series with respect to method B; while $C \nrightarrow D$ means there is a series converging with respect to method C and diverging with respect to method D. These introductory remarks already show a separation between the one and two dimensional situations—things will usu-

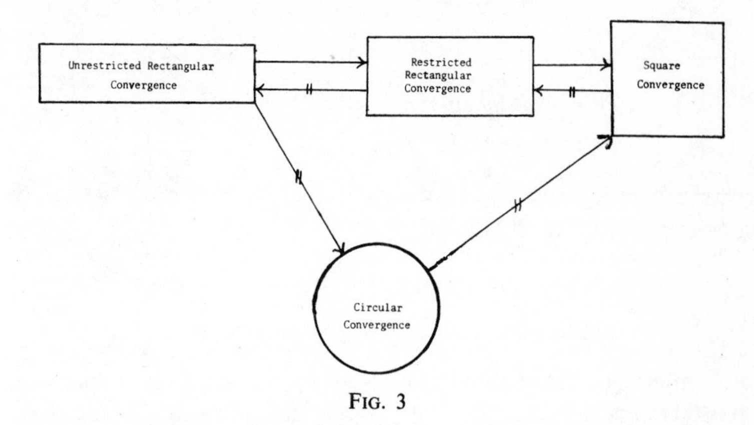

FIG. 3

ally remain worse for the two dimensional case than for the one dimensional case when we throw some exponentials in and begin our study of trigonometric series.

The main body of my discussion will be broken into five areas. They will be convergence and divergence of Fourier series, the effect of convergence on the size of coefficients, localization, the relative behaviors of a trigonometric series and its conjugates, and the uniqueness of the representation by trigonometric series. In each area, I will very quickly run over the main points of the one dimensional theory, and then tell you how far the results can be extended into higher dimensions.

2. CONVERGENCE AND DIVERGENCE OF TRIGONOMETRIC SERIES

Here the one dimensional theory is pretty clear. The positive result is the

CARLESON-HUNT THEOREM: *If $f \in L^p(T)$, $p > 1$, then $S_n[f](x)$ converges at almost every point x.* (2.1)

Carleson did the $p = 2$ case in 1965 and Hunt extended to all $p > 1$ in 1967 [4], [12]. I want to mention two negative results. The first is the very well-known counterexample of A. N.

Kolmogorov:

There is a function k in $L^1(T)$ whose Fourier series diverges everywhere. (2.2)

The other negative result, which is due to J. Marcinkiewicz, is possibly not quite as familiar. It is

There is a function $m \in L^1(T)$ which is finitely oscillating and divergent almost everywhere [18]. (2.3)

Finitely oscillating means that if we fix a point x and look at the partial sums $s_n = S_n[f](x)$, they wiggle (i.e., $\limsup s_n - \liminf s_n > 0$) but they do not go off to infinity (i.e., there is a finite number $M(x)$ such that $\sup |s_n| \leqslant M$).

Kolmogorov in 1923 produced a weaker counterexample of an L^1 function with series divergent only almost everywhere [15]. (This means that there was an exceptional set of zero Lebesgue measure on which convergence might occur.) Three years later, he was able to construct the perfectly divergent function of (2.2) above [16]. We emphasize this point because the "almost" of (2.3) cannot be dropped. For suppose there were an $L^1(T)$ function $m_1(x)$ finitely oscillating at each x. Then the union of the closed sets $E_N = \{x \in T : \sup_n |S_n[m_1](x)| \leqslant N\}$ as N ranges over the positive integers would be all of T, so that by the Baire category theorem some E_{N_0} would contain an open interval (a, b) in which all the $S_n[m_1](x)$ and so also $m_1(x)$ would be bounded by N_0. (Recall m_1 is the $(C, 1)$ limit of its partial sums almost everywhere.) Hence replacing m_1 by 0 outside (a, b), we would obtain a new function m_2 bounded on T and thus by (2.1) convergent almost everywhere on T. But $m_1 \equiv m_2$ on (a, b) so that by localization (see section 4 below) m_2 must diverge almost everywhere on (a, b)—a contradiction.

In two dimensions, the answer to the question of convergence varies with the method of summation. To start with let's look at restricted rectangular convergence. There is a tremendously good function f with an everywhere divergent Fourier series. How good is f? This good:

(i) f is continuous,
(ii) f has a Fourier series of power series type, (2.4)
(iii) f has everywhere uniformly bounded rectangular partial sums.

Condition (i) is much stronger than $f \in L^p$, so we have a marked contrast to (2.1). By condition (ii) we mean that in f's Fourier expansion (see 1.1) the coefficients $\hat{f}_{mn}$ are zero if (m, n) is not in the first quadrant. The one dimensional analogue of this—$\hat{f}_m = 0$ if $m < 0$—often improves things greatly; it doesn't seem to help much here. The impossibility of dropping the "almost" from the conclusion in Marcinkiewicz's example (2.3) shows that condition (iii) is in very dramatic contrast to the one dimensional case. If, however, you prefer your examples more divergent, it is easy to change f so that (iii) is replaced by "f has $\limsup_{n\to\infty}|S_n[f](x)| = \infty$ everywhere". This example is essentially due to Fefferman in 1970 [10] with a few of the frills added in [1].

You can get positive results if you shift gears by changing the method of convergence. Recall from our discussion of numerical series that it would seem easier for a series to be square convergent; and, sure enough, it is a lot easier. Here is a theorem which indicates this:

If $f \in L^p$, $p > 1$, then $S_n[f] \to f$ almost everywhere. (2.5)

This was proved around 1970 simultaneously by Fefferman (in the United States), Sjolin (in Sweden), and Tevzadze (in the USSR) [9], [22], [25]. The method of proof is quite interesting. We would like to do induction starting from Carleson's theorem (2.1) but for technical reasons this doesn't work. So we introduce a "butterfly". If

$$f \sim \sum \sum \hat{f}_{mn} e^{2\pi i(mx+ny)},$$

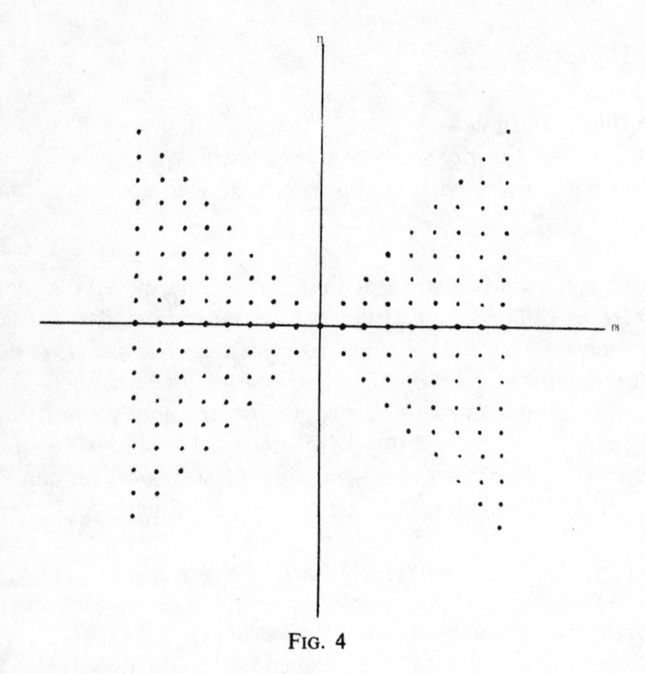

FIG. 4

decompose the sum into

$$\sum_{m=-\infty}^{\infty} \sum_{n=-|m|}^{|m|} + \sum_{m=-\infty}^{\infty} \sum_{|n|>|m|} . \tag{2.6}$$

To both terms correspond functions which are also in L^p (M. Riesz's theorem) and for both these functions the lattice plane regions being summed over are now butterflies (e.g., the first summand's index set looks like Figure 4), and the technical obstacle to the induction vanishes, as is easily seen.

I will mention in passing that for double Walsh-Fourier series it is not clear if both of the sums in (2.6) represent L^p functions if $p \neq 2$, so the analogue of theorem (2.5) for double Walsh series follows easily only for $p = 2$. If $p < 2$, the question remains open.

For circular convergence there are some negative results but there is still an open question. Here if $f \in L^p$ with $p < 2$, circular divergence may occur. This follows from Fefferman's counterexample for the multiplier problem [7], [8]. If $p = 2$, the question remains open. I think this is a nice question to work on.

There are various ways to clean up the counterexample (2.4). One thing we can do is to put more stringent restrictions on the function f. For example, we might demand that $f \in L^2(T^2)$ and that one of its partial derivatives exist and also belong to $L^2(T^2)$ [22]. A second approach is to work with summability instead of convergence. For example, for f in L^p the $(C, 1, 0)$ means of the Fourier series—

$$\sigma_{mn} = \frac{1}{m+1}(S_{0n} + S_{1n} + \cdots + S_{mn})$$

do converge to f unrestrictedly rectangularly almost everywhere [3]. (We were able to get away with averaging only in the m direction because of Carleson's theorem.)

3. CONVERGENCE AND GROWTH OF COEFFICIENTS

The point here will be to see how bad coefficients can be for a convergent trigonometric series.

The one dimensional situation is very nice. We assume that $T = \Sigma a_n e^{2\pi inx}$ converges on the set E, where $|E|$—the Lebesgue measure of E—is greater than 0. By definition this means that

$$a_n e^{2\pi inx} + a_{-n} e^{-2\pi inx} \to 0, \qquad x \in E, |E| > 0. \tag{3.1}$$

The Cantor-Lebesgue theorem postulates (3.1) and concludes that

$$a_n \to 0 \quad \text{as} \quad |n| \to \infty.$$

So convergence on a set of positive measure forces the coefficients to go to 0. Again, the answer as to whether this works in higher dimensions depends on the method of convergence. We will assume throughout this section that a double trigonometric series $T = \Sigma a_{mn} e^{2\pi i(mx+ny)}$ is convergent (under the mode of convergence being discussed) at each $x \in E$, where $|E| > 0$.

For circular convergence the conclusion is very nice. Not only do the coefficients go to zero, but, in fact, the sum of the squares of the coefficients lying on a circle

$$\sum_{\{(m,n):m^2+n^2=r^2\}} |a_{mn}|^2$$

tends to 0 as the radius r tends to ∞. This was first proved when E has full measure by Roger Cooke in 1971 and the hypothesis of $|E| = 1$ was weakened to $|E| > 0$ by Antoni Zygmund in 1972 [6], [28]. The partial sums are no problem because we are dealing with a one-parameter family. For one parameter methods (e.g., circular —which is indexed by r—and square—which is indexed by (n, n)) convergence implies that the partial sums tend to a limit by definition.

Let us now pass to the unrestricted rectangular case. Here convergence doesn't quite force the coefficients to go to 0. It does force

$$a_{mn} \to 0 \text{ in the northeast.}$$

We can prove this as fast as we can explain what is meant. Fix $(x, y) \in E$ and take a rectangular partial sum $S_{m,n} = S_{m,n}(x, y)$ which is close to the limit (this will be true if m and n are both big); add $S_{m-1,n-1}$; then subtract $S_{m-1,n}$ and $S_{m,n-1}$. All 4 of these partial sums are near the limit, so since 2 are taken positive and 2 are taken negative the resultant expression is small. But a moment's thought will show that this expression has all terms 0 except for the (m, n) one:

$$a_{m,n}e^{2\pi i(mx+ny)} + a_{m,-n}e^{2\pi i(mx-ny)} + a_{-m,n}e^{2\pi i(-mx+ny)} + a_{-m,-n}e^{-2\pi i(mx+ny)}. \tag{3.2}$$

This argument, which is nothing more than the two dimensional version of $a_n = \Sigma_{\nu=1}^{n} a_\nu - \Sigma_{\nu=1}^{n-1} a_\nu$, is what we call the Mondrian proof. (See [3, p. 411] for a picture and further details.) Finally, there is an easy inductive extension of the Cantor-Lebesgue theorem that deduces from (3.2) that the coefficients themselves must be small. But notice that the reasoning that made (3.2) small required *both* m and n to be large—that is (m, n) to be in the northeast.

Example 4 above might seem to contradict this proof. What's going on, of course, is that the terms are not in the northeast; rather they are all due east—that is, down on the m axis. It looks as though you have no control at all over what happens in the east. However, it turns out that using a very clever lemma of Paul J. Cohen, you can prove that the coefficients are all bounded regardless of where they are:

$$|a_{mn}| \leqslant M \text{ for some } M \text{ and all } (m, n),$$

[3, p. 410]. If we keep thinking about numerical example 4 we can see that this is a little bit surprising. Near the end of section 1, we said that double trigonometric series were usually just as bad as double numerical series. Here is one case where they are better. We can use the positive measure of the set E to disallow possibilities like those of example 4.

Square convergence is awful. The series can square converge on a set of positive measure, even on a set of full measure and still have incredibly big coefficients. Here is the example:

$$T = \sum_{n=1}^{\infty} n^{10^6} (\sin^2 \pi x)^n e^{2\pi i n y}. \tag{3.3}$$

This looks like a single series—we can see an x and a y in it but there is only one summation sign. To see what's going on, I'll change the sines into exponentials via Euler's formula:

$$T = \sum \frac{n^{10^6}}{(2i)^{2n}} \left(e^{\pi i x} - e^{-\pi i x}\right)^{2n} e^{2\pi i n y}.$$

If we apply the binomial theorem, we find

$$\begin{aligned}\left(e^{\pi i x} - e^{-\pi i x}\right)^{2n} = e^{2\pi i n x} + \cdots + (-1)^n \binom{2n}{n} \\ + \cdots + e^{-2\pi i n x}\end{aligned} \tag{3.4}$$

so that the n-th term of the summation actually includes terms associated to the lattice points on the "up" butterfly wing at height n:

$$(-n, n), (-n+1, n), \ldots, (0, n), \ldots, (n, n).$$

Thus the partial sums of the single series (3.3) are exactly the square partial sums of T thought of as a double trigonometric series; so if we want to know whether T is square convergent we simply have to ask whether the single series (3.3) is convergent. Well you can tell we're up to tricks with the n^{10^6} term. On the other hand the size of the n-th term is being driven down much faster, in fact geometrically by the $(\sin^2 \pi x)^n$ term. We're in trouble only if $\sin^2 \pi x = 1$ and this occurs only on the extreme edge of T^2, that is, only on the line $x = -\frac{1}{2}$. So T is convergent on $(-\frac{1}{2}, \frac{1}{2}) \times [-\frac{1}{2}, \frac{1}{2})$, which is certainly almost everywhere. Now how about the coefficients? This sounds like a horribly messy job, but it's not. In fact, we'll take only one term from the expansion (3.4)—the middle term $(-1)^n\binom{2n}{n}$—which corresponds to the coefficient $a_{0,n}$. To avoid worrying about i's and minus signs we'll look only at absolute values. We have

$$|a_{n,0}| = \frac{n^{10^6}}{2^{2n}}\binom{2n}{n}.$$

From Stirling's formula it easily follows that

$$\frac{1}{2^{2n}}\binom{2n}{n} \cong \frac{1}{\sqrt{\pi n}}$$

so that

$$|a_{n,0}| \cong \frac{1}{\sqrt{\pi}} n^{999{,}999.5},$$

which is quite unbounded; coefficients of square convergent series can be pretty bad. (See [3, p. 408] for details.)

Recall the diagram at the end of the introduction which summarized the incompatibility of the four convergence methods for numerical series. Examples such as 3.3 allow us to construct a similar diagram in which, for example, "$A \nRightarrow B$" means "There is a trigonometric series T converging at each point of some set E of positive measure by method A, but not converging at any point of E by method B". All the arrows (and non-arrows) are the same except that it *may* be that the circular convergence of T on a set forces the square (or perhaps even the unrestricted rectangular) convergence of T at almost every point of that set. This is an open question [3, p. 420].

It is interesting to compare the results of sections 2 and 3 for the various methods of summation. In section 2 we wanted a function's goodness to force its Fourier series convergence. Since it is very easy for series to square converge, in section 2 one gets the best theorem for square convergence—in fact one gets essentially no theorem at all for the other rectangular methods. Conversely in section 3 the assumption is convergence and the hoped for conclusion is good behavior of the series coefficients. Here since unrestricted rectangular convergence is the most difficult of the rectangular methods, it provides the strongest hypothesis and hence the best theorems, while square convergence yields the poorest conclusions. In short, which method of summation is best depends on what you're trying to do.

4. LOCALIZATION

We again start with the one dimensional case, letting our function f belong to $L^1(T^1)$. If $f = 0$ in a neighborhood of the point x; then, regardless of how bad f is anywhere else, its Fourier series converges to 0 at x. This phenomenon is called localization. In other words, the behavior of the partial sums of the Fourier series only depends on how the function looks right near the point.

Now for 2 variables and rectangular methods, localization fails. In fact, there is an f with a differential at each point (so that in particular at each (x, y) $\partial f/\partial x$ and $\partial f/\partial y$ exist), a point (x_0, y_0), and neighborhood N of (x_0, y_0) on which $f \equiv 0$, for which even the square partial sums (and hence *a fortiori* the other two types) get out of hand:

$$\sup_n |S_{n,n}[f](x_0, y_0)| = \infty.$$

For circular convergence localization fails again. I'm not sure if you can do it with a differentiable function (I wouldn't be surprised if you could), but I know it fails with a continuous function [13], [14].

So localization is a complete failure—the exact analogue of the one dimensional result is simply false.

Well, as usual there are a lot of ways around the problem—at least three. One is to demand that f be very, very smooth. For example, we can demand that the two partial derivatives be themselves L^1 functions—i.e., that f belong to the so-called Sobo-

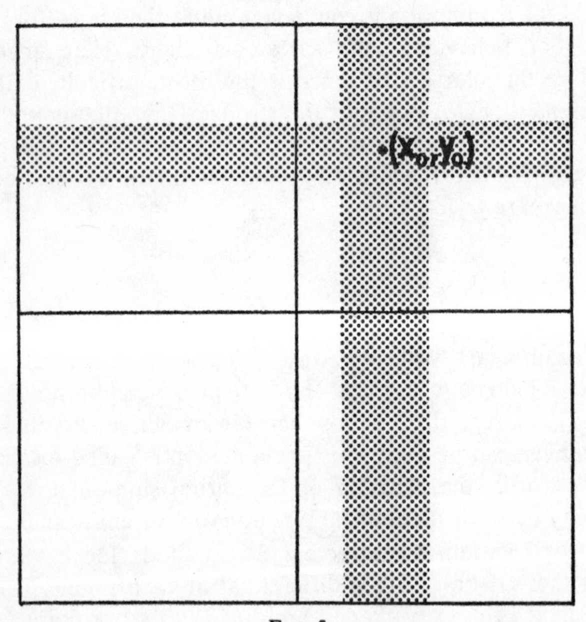

FIG. 5

lov space $W_{1,1}$. There is a 1972 work of Goffman and Liu [11] and another paper of Liu (also in 1972 [17]) on this.

A second way out is to make f more zero. If it's not 0 enough already, make it more 0—make it 0 on an entire cross-neighborhood of (x_0, y_0). We won't define a cross-neighborhood but it looks like Figure 5. The cross may be very thin but it has to go all the way to the boundary of T^2 in all 4 directions. If $f = 0$ on such a neighborhood, then you will get localization for all the rectangular methods.

A third way is to make f a little bit good—say continuous—and to replace convergence by summability. For example, the $(C, 1, 1)$ partial sums

$$\frac{1}{m+1}\,\frac{1}{n+1}\sum_{i=0}^{m}\sum_{j=0}^{n} S_{ij}[f](x_0, y_0)$$

will converge to 0 when f is 0 on an (ordinary) neighborhood of (x_0, y_0) [27, vol 2, p. 305]. There are similar results for circular summability [24].

5. PLESSNER'S THEOREM

Again in one dimension, this is fun, easy to state, and pretty, although somewhat hard to prove. We look at a trigonometric series

$$T = \sum_{n=-\infty}^{\infty} a_n e^{2\pi i n x}$$

and we look at its positive half

$$T^+ = \sum_{n=0}^{\infty} a_n e^{2\pi i n x}.$$

If this general trigonometric series T converges on a set E of positive measure, then T^+ will also converge on the same set up to a set of measure 0. Equivalently, we can look at the trigonometric series conjugate to T

$$\tilde{T} = -\sum_{n=-\infty}^{\infty} (i \operatorname{sgn} n) a_n e^{2\pi i n x}, \qquad \operatorname{sgn} n = \begin{cases} 1, & n > 0 \\ 0, & n = 0 \\ -1, & n < 0 \end{cases}.$$

If we look at T, T^+, and $\tilde{T}$ we can see simple algebraic relationships between them (such as $T + i\tilde{T} = 2T^+$) from which it immediately follows that the connection between T and $\tilde{T}$ is the same as that between T and T^+. So another equivalent statement of Plessner's theorem is that if T converges on E, then $\tilde{T}$ converges almost everywhere on E [27, vol 2, p. 216].

In two variables we start with a double series

$$T = \sum_{m=-\infty}^{\infty} \sum_{n=-\infty}^{\infty} a_{mn} e^{2\pi i(mx+ny)}.$$

Now, what are our analogues of T^+ and $\tilde{T}$ going to be? Well by T^+ we might mean

$$T^{+x} = \sum_{m=0}^{\infty} \sum_{n=-\infty}^{\infty} a_{mn} e^{2\pi i(mx+ny)}$$

(this is chopping off half the lattice plane) or we might mean

$$T^{++} = \sum_{m=0}^{\infty} \sum_{n=0}^{\infty} a_{mn} e^{2\pi i(mx+ny)}$$

(which amounts to just grabbing a quarter of the lattice plane).

For all four methods of convergence, these extensions of Plessner's theorem are false. In fact, we can actually let T be the Fourier series of an $L^1(T^2)$ function and get it to converge on a set E, $|E| > 0$, and get both T^{+x} and T^{++} to diverge almost everywhere on E [2]. (This may clarify my introductory remark about the lack of naturality in grouping the (m, n), $(m, -n)$, $(-m, n)$ and $(-m, -n)$ terms of a series together.)

Again we can salvage something here in a variety of ways. One way is to improve the series to be actually the Fourier series of a function in $L^p(T^2)$ for $p > 1$. Then the theorem will be true for unrestricted rectangular convergence but only in two dimensions. (Not in 3 or more [2].) That's one way—improve T. Or, we can add additional technical hypotheses such as the following: If we assume that T^{+x} is a little bit good *and* that T converges on a set E, then we can make it. (For example, that T^{+x} be summable $(C, 1, 0)$ [2].) Or, another way out of the bag is to use some other notion of conjugate altogether [19].

6. UNIQUENESS

The last topic is the best, or the worst, because here almost all the good theorems are questions. Professor Zygmund's article points out that the problem of uniqueness in one variable is a long way from over because of all the difficulty with sets of uniqueness. Well, in several variables it's even further from being over because we haven't even been able to resolve the higher dimensional analogues of the primitive, simple-minded, first draft version theorem due to Riemann which says that if a trigonometric series converges to 0 at every $x \in T^1$, then actually it isn't there—all its coefficients were 0. That's the beginning of the subject and we can't even do that, in several variables, very often. So

$$\textit{if } T_n(x) \to 0 \textit{ for every } x, \textit{ then all coefficients are } 0, \tag{6.1}$$

is the very pretty one dimensional theorem. It's very simple and nice—we just integrate twice formally and observe that the twice integrated series has second generalized derivative equal to 0, so the twice integrated series is convex, concave, and hence linear, and therefore the original series was 0. This is a very clean argument, but the only known argument. There is no other known way to do it, so far. It would be really nice to have another way of doing this.

In two dimensions we have a couple of positive results; and they're only good for two dimensions. One is

$$\textit{If } T_{mn} \to 0 \textit{ unrestrictedly rectangularly everywhere, then all } a_{mn} = 0. \tag{6.2}$$

The other one is

$$\text{If } T_r \to 0 \text{ circularly everywhere, then all } a_{mn} = 0. \qquad (6.3)$$

Basic to both of these proofs is some work of Victor Shapiro involving uniqueness for spherical Abel summability—a very nice paper [21]. Roger Cooke used his fact about coefficients tending to 0 (that I mentioned earlier) together with Shapiro's theorem (which has an assumption concerning coefficient size) to obtain (6.3) [6]. Theorem (6.2) was obtained *also* using Shapiro's theorem which is really kind of crazy when you think about it—to use something about circular means to end up with a rectangular result [3, p. 422]. Examples 1 and 4 seem to emphasize this point. Nevertheless Theorem (6.2) cannot be done directly by rectangular methods—not by me, anyway. That's the end of the positive results in several variables. Notice I was very careful to say "two dimensions" in those two theorems. So that leaves just about every other question in the field wide open. For example, what if you assume that the series is still two dimensional, but converges only *restrictedly* rectangularly to 0 everywhere—what then? And what if you look at a triple trigonometric series which converges to 0 everywhere and you name the method of convergence—what then?

REFERENCES

1. Ash, J. M. and L. Gluck, "A divergent multiple Fourier series of power series type," *Studia Math.*, **44** (1972), 477–491.
2. ——, , "Convergence and divergence of series conjugate to a convergent multiple Fourier series," *Trans. Amer. Math. Soc.*, **207** (1975), 127–142.
3. Ash, J. M., and G. V. Welland, "Convergence, uniqueness, and summability of multiple trigonometric series," *Trans. Amer. Math. Soc.*, **163** (1972), 401–436. MR45#9057.
4. Carleson, L., "On convergence and growth of partial sums of Fourier series," *Acta Math.*, **116** (1966), 135–157. MR33#7774.

5. Chandrasekharan, K., "On some problems in Fourier analysis", *Proc. International Congress of Mathematicians*, 1954, Amsterdam, Vol. III, pp. 85–91. Erven P. Noordhoff N.V., Groningen, North Holland Pub. Co., Amsterdam, 1956.

6. Cooke, R., "A Cantor-Lebesgue theorem in two dimensions," *Proc. Amer. Math. Soc.*, **30** (1971), 547–550. MR43#7847.

7. Fefferman, C., "Inequalities for strongly singular convolution operators," *Acta. Math.*, **124** (1970), 9–36. MR41#2468.

8. ——, , "The multiplier problem for the ball," *Ann. of Math.*, (2) **94** (1971), 330–336. MR45#5661.

9. ——, , "On the convergence of Fourier series," *Bull. Amer. Math. Soc.*, **77** (1971), 744–745.

10. ——, , "On the divergence of Fourier series," *Bull. Amer. Math. Soc.*, **77** (1971), 191–195. MR 43#5251.

11. Goffman, C., and F. Liu, "On the localization property of square partial sums for multiple Fourier series," *Studia Math.*, **44** (1972), 61–69. MR47#709.

12. Hunt, R. A., "On the convergence of Fourier series," *Proc. Conf. Orthogonal Expansions and their Continuous Analogues* (Edwardsville, Ill., 1967), Southern Illinois Univ. Press, Carbondale, Ill., 1968, 235–255. MR38#6296.

13. Igari, S., "On the localization property of multiple Fourier series," *J. Approximation Theory*, **1** (1968), 182–188. MR40#641.

14. ——, , "Lectures on Fourier series of several variables," Univ. of Wisconsin Lecture Notes, Madison, Wis., (1968).

15. Kolmogorov, A. N., "Une série de Fourier-Lebesgue divergente presque partout," *Fund. Math.*, **4** (1923), 324–328.

16. ——, , "Une série de Fourier-Lebesgue divergente partout," *Comptes Rendus*, **183** (1926), 1327–8.

17. Liu, F., "On the localization of rectangular partial sums for multiple Fourier series," *Proc. Amer. Math. Soc.*, **34** (1972), 90–96. MR45#4061.

18. Marcinkiewicz, J., "Sur les séries de Fourier," *Fund. Math.*, **27** (1936), 38–69.

19. Peterson, G. E., and G. V. Welland, "Plessner's theorem for Riesz conjugates," *Pacific J. Math.*, **60** (1975).

20. Shapiro, V. L., "Fourier series in several variables," *Bull. Amer. Math. Soc.*, **70** (1964), 48–93. MR28#1448.

21. ——, , "Uniqueness of multiple trigonometric series," *Ann. of Math.*, (2) **66** (1957), 467–480. MR 19, 854; 1432.

22. Sjolin, P., "On the convergence almost everywhere of certain singular integrals and multiple Fourier series," *Ark. Mat.*, **9** (1971), 65–90.

23. ——, , "On the convergence almost everywhere of double Fourier series," Uppsala Univ., Dept. of Math., Report No. 3 (1975), 1–9.

24. Stein, E. M., "Localization and summability of multiple Fourier series," *Acta Math.*, **100** (1958), 93–147. MR21#4331.

25. Tevzadze, N. R., "On the convergence of double Fourier series of quadratic summable functions," *Soobšč. Akad. Nauk Gruzin. SSR.*, **5** (1970), 277–279.

26. Zhizhiashvili, L. V., "Some problems in the theory of simple and multiple trigonometric and orthogonal series," *Russian Math. Surveys*, **28** (1973), 65–127.

27. Zygmund, A., *Trigonometric Series*, Vols. 1, 2, 2nd rev. ed., Cambridge Univ. Press, New York, 1959. MR21#6498.

28. ——, , "A Cantor-Lebesgue theorem for double trigonometric series," *Studia Math.*, **43** (1972), 173–178. MR47#711.

For more extensive discussion of the field as a whole and bibliographies, the reader is encouraged to seek out [5], [20], [26], or [14]. The first two of these are mainly about spherical methods with which I deal only lightly. The third has a mammoth bibliography, while the last surveys a lot of topics I do, giving proofs.

HARMONIC ANALYSIS ON R^n

Elias M. Stein

What I will try to do in this article,* and probably not succeed at at all, is to give you a mixture of two things—first recalling very quickly some elementary and standard things about harmonic analysis on Euclidean spaces and at the same time mixing this in with some rather recent results—some new results on the frontier of research, and second recalling some interesting problems that these raised. You can imagine that this might be difficult to do.† There probably won't be enough time to do the things that I want to do.

---

*As the attentive reader might guess, what follows is a rather faithful transcription (with the aid of videotape) of the actual two hours of lectures as given. I gratefully acknowledge the generous help of Marshall Ash for making this possible.

†This survey must, by its circumstances, be very incomplete. Thus the desire to given an "elementary" presentation makes it impossible to describe some important ideas which by their nature are "technical". Moreover, the wish not to arrogate for myself the topics of other lecturers leads me to say little about such closely related areas as H^p spaces, analysis on homogeneous and symmetric spaces, etc.

1. INTRODUCTION

Let me begin now. I am going to divide this article into a discussion of three topics. Here is an outline of where I will be going. The first topic will of course be the Fourier transform. The second topic will be some discussion about convolution operators; and the third topic, again related to these two, will be highlighting certain symmetry properties of R^n. And all of these topics are going to be intermingled.

2. THE FOURIER TRANSFORM

Let's first of all come to the Fourier transform. I've decided that I'm going to assume that you know its elementary properties [6, p. 1]. On R^n, that is Euclidean n-space, we're going to be talking about the Fourier transform and the definition of the Fourier transform is the following integral

$$\int_{R^n} e^{2\pi i x \cdot y} f(y)\, dy.$$

Here I use that normalization with 2π in the exponential—there's always some normalization with ± 1 or $\pm 2\pi$ entering, x and y are vectors in R^n, $x \cdot y$ is the standard inner product in R^n, f is a given function and dy is the standard Euclidean measure on R^n. Now that is the definition of the Fourier transform. It makes sense for appropriate functions f. We denote it by $\hat{f}(x)$, so that $\hat{f}(x) = \int_{R^n} e^{2\pi i x \cdot y} f(y)\, dy$. The basic properties about the Fourier transform are, among other things, that for appropriate functions f the function $\hat{f}$ is of course well defined, and you can recover f from $\hat{f}$ by a similar transformation, namely the inverse Fourier transform which I will now write down,

$$f(y) = \int_{R^n} e^{-2\pi i x \cdot y} \hat{f}(x) dx.$$

I am writing this in a purely formal way. Nothing has been said yet about questions of convergence. Now the first thing that occurs at the beginning is a class of functions that could be used for which this inversion is rigorous. A well-known class which is very useful by now is a class of testing functions of Schwartz. This class of testing functions of Schwartz is a subset of the class of all functions on R^n which are infinitely differentiable with all their partial derivatives continuous. But we require not only that the functions be infinitely differentiable, but that they be suitably small at infinity, together with all their derivatives. That means that if we take $\partial_x^\alpha f$, the x derivative of f of order α ($\partial_x^\alpha f$ is a shorthand notation for

$$\left(\frac{\partial}{\partial x_1}\right)^{\alpha_1}\left(\frac{\partial}{\partial x_2}\right)^{\alpha_2}\cdots\left(\frac{\partial}{\partial x_n}\right)^{\alpha_n} f$$

where x_1 through x_n are the coordinates of the point x, α_1 through α_n are non-negative integers—this is the α-th differential monomial), and multiply by any monomial in x, let's say x^β (again, $x^\beta = x_1^{\beta_1}x_2^{\beta_2}\cdots x_n^{\beta_n}$) and require that the resultant function $x^\beta\partial_x^\alpha f$ be bounded, we arrive at the Schwartz class. Notice that we may apply ∂_x^α since f is assumed to be infinitely differentiable. In short, $\mathcal{S} = \{f \in C^\infty |\, |x^\beta \partial_x^\alpha f| \leqslant M < \infty \forall \alpha, \beta \geqslant 0\}$ is the class of testing functions of Schwartz. Now a basic property of these testing functions is that if f is such a testing function, it is easy to verify that $\hat{f}$ is such a testing function. The basic reason for that is that the Fourier transform of course has the property that it takes, except for factors of $2\pi i$, differentiation into multiplication by the corresponding monomial and vice versa.

$$\partial_x^\alpha \hat{f}\,(x) = ((+2\pi i)^{\alpha_1 + \cdots + \alpha_n} y^\alpha f)\hat{}\,(y),$$

$$(y^\alpha f)\hat{}\,(x) = (-2\pi i)^{\alpha_1 + \cdots + \alpha_n} \partial_x^\alpha \hat{f}\,(x)$$

see [13, p. 5], but notice that the slightly different definition there switches the minus signs around.) It only takes a little bit of extra reasoning to show that the Fourier transform maps the space $\mathcal{S}$ into itself, and therefore one can now state the Fourier inversion formula at least on the space $\mathcal{S}$. That is to say, if f is in $\mathcal{S}$, then $\hat{f}$ is in $\mathcal{S}$ and when we take the inverse Fourier transform, which is the same as the original Fourier transform except for the change of sign in the exponent, then we get back the function-$f(y) = \int_{R^n} e^{-2\pi i x \cdot y} \hat{f}(x) dx$. That is a simple remark about the Fourier inversion formula. Another basic remark is the Plancherel theorem. One part of the Plancherel theorem is Parseval's identity

$$\int_{R^n} |f|^2 dx = \int_{R^n} |\hat{f}|^2 dx.$$

This is still for functions f in $\mathcal{S}$, but notice that functions which are in the testing space of Schwartz are in L^2, so again everything is well defined. This together with the inversion formula leads to the Plancherel theorem which says that the Fourier transform, initially defined on the subspace $\mathcal{S}$, has a unique extension to a unitary operator on all of $L^2(R^n)$. (Of course $L^2(R^n)$ is the space that arises by taking the testing functions $\mathcal{S}$, using as a norm $\left(\int_{R^n} |f|^2 dx\right)^{1/2}$, and taking the closure in that norm.) That is the Plancherel theorem. And I would also state as a theorem, the Fourier inversion theorem, at least for testing functions. These are some very simple properties of the Fourier transform.

Where do the symmetry properties of R^n enter? First of all, R^n is an abelian group under vector addition. Thus, there is of course the action of translations given by the mapping $x \to x - h$, where h in R^n is fixed. Then there is the induced action on functions, $f(x) \to f(x - h)$. Notice that this translation-induced action preserves $\mathcal{S}$, the class of testing functions; and it preserves $L^2(R^n)$ also. Now the Fourier transform behaves in a very simple manner with respect to translations. Because if you translate the function f first and then take the Fourier transform, or if you take the Fourier transform first and then multiply by $e^{2\pi i x \cdot h}$, you get the same effect:

$$\begin{array}{ccc} f(y) & \to & f(y-h) \\ \downarrow & & \downarrow \\ \hat{f}(x) & \to & \hat{f}(x) e^{2\pi i x \cdot h} \end{array}.$$

So the Fourier transform behaves in a very simple way on the translations. You can see that merely by making a change of variables in the integral $\int e^{2\pi i x\cdot y}f(y-h)dy$. And one can say that the real role, in some sense, of the Fourier transform is the following. On the $L^2(R^n)$ space, which is a Hilbert space, consider the family of all translations. That is a commutative family of unitary operators. The effect of the Fourier transform is in some sense to simultaneously diagonalize all these unitary operators, that is to give them by the multiplication operators $e^{2\pi i x\cdot h}$. So the Fourier transform arises, you might say, because of the translation structure of R^n. That's one of the symmetry properties that R^n has.

But it also has some other symmetry properties which we will return to later on. I shall just mention them briefly. What else can you do to R^n besides translating? One of the things you can do is to carry out a stretching. A stretching is of course extremely important: I'll call it a *dilation*. And that's a mapping which sends $x \to \delta x$, where the real number δ is greater than 0. These dilations have an important effect on Fourier analysis. If we replace $f(y)$ by $f(\delta y)$ and then take the Fourier transform, the Fourier transform of the dilated function also is closely related to $\hat{f}(x)$. What you are led to is the inverse dilation, multiplied by δ to the power $-n$, where n is the number of dimensions:

$$\begin{array}{ccc} f(y) & \rightarrow & f(\delta y) \\ \downarrow & & \downarrow \\ \hat{f}(x) & & \delta^{-n}\hat{f}(\delta^{-1}x) \end{array}$$

So the Fourier transform transforms in a very nice natural way under dilations.

Finally, there is a last property which is very important for us, and that is the effect of rotations of the underlying space. The basic fact here is that the Fourier transform actually commutes with rotations. (Throughout this paper, "rotation" will always mean "rotation about the origin".) If I rotate the space and then take the Fourier transform, that is the same as if I took the Fourier transform first and then rotated. The reason for this is very simple.

It is that the Fourier transform is defined in terms of the inner product $x \cdot y$ and of course the inner product is a basic two-point invariant under rotation. So, in some sense, anything that can be said about Euclidean Fourier analysis will have to reflect this symmetry structure. There will be a little bit about this later on.

3. FORMULAS FOR FOURIER TRANSFORMS

Now I would like to come to some formulas. Analysis, as you all know, is a subject replete with inequalities and estimates of all kinds. But like any other subject of mathematics, there are at its basis some fundamental identities, formulas which are at once explicit, far-reaching and elegant. I would like to mention a few of them because they will be important later on. The formulas I want to mention are three. They are all related to each other. The first formula is the following: Suppose I were to take the function $f(y) = e^{-\pi|y|^2}$. That is the Gaussian function, the Gaussian exponential with the right normalization (because of our particular definition of the Fourier transform). The remarkable fact is that the function is its own Fourier transform:

$$\hat{f}(x) = e^{-\pi|x|^2}. \tag{1}$$

How do you prove this formula? Well, this formula is clearly independent of the dimension n since

$$\int_{R^n} e^{-\pi|y|^2} e^{2\pi i x \cdot y} dy = \int_{-\infty}^{\infty} e^{-\pi y_1^2 + 2\pi i x_1 y_1} dy_1$$

$$\int_{-\infty}^{\infty} e^{-\pi y_2^2 + 2\pi i x_2 y_2} dy_2 \cdots \int_{-\infty}^{\infty} e^{-\pi y_n^2 + 2\pi i x_n y_n} dy_n.$$

Thus because of this product nature of the being, it's just a question of proving it in one variable. This is effected by a very well-known contour integration [6, p. 43]. That is one formula; it is the starting point for many other formulas.

The second formula is a consequence of the statement that we just made about rotations. We observed that if one takes a function f and rotates the underlying variable and then takes the Fourier transform, that's the same as if one just took the Fourier transform and then rotated the variable for the function $\hat{f}$. Now let us consider functions which are independent of rotations. That is, suppose f is a *radial* function—a function which is independent of rotation—a function which depends only on distance from the origin. So such a function $f(x)$ can be written as another function of $|x|$, to which (by abuse of language) we give the same name, $f(x) = f(|x|) = f(r)$. Now the Fourier transform of such a function—this is now a function of one variable disguised as a function of n variables—is another n variable function, which is again really a one variable function. Now there is an explicit formula for this, and the explicit formula is

$$\hat{f}(r) = 2\pi r^{-(n-2)/2} \int_0^\infty f(s) J_{(n-2)/2}(2\pi r s) s^{n/2}\, ds. \tag{2}$$

The function J is a Bessel function. This is where the theory of Bessel functions enters. So, if you start with a radial function, the Fourier transform of it is again another radial function and the induced action is given in terms of a Bessel function of order $(n-2)/2$. Therefore, one of the important consequences is that the theory of the Bessel functions is intimately connected with n-dimensional Fourier analysis. I'm not going to write down, however, the definition of the Bessel function except to remind you of some very special cases which occur. Let us take the simplest case of all, which is $n = 1$, that is the real line. The radial functions now are the even functions. This Bessel function $J_{(n-2)/2}$ is now a Bessel function of order $-\frac{1}{2}$. For all intents and purposes the Bessel function $J_{-1/2}$ is the cosine function. (Precisely, $J_{-1/2}(x) = \sqrt{2/\pi}\,(\cos x)/\sqrt{x}$.) And so the Fourier transform, which of course is an even function, is a cosine transform—that's an elementary thing. So matters are quite elementary when $n = 1$. Matters are also more or less elementary when $n = 3$. Because in this case we are talking about the Bessel function of order $\frac{1}{2}$ and that's the sine function and that is also a rather elementary transform. (Precisely, $J_{1/2}(x) = \sqrt{2/\pi}\,(\sin x)/\sqrt{x}$.) But when n is equal to 2 matters are not so elementary, because we get a Bessel function of order 0, and that is not an elementary function; it cannot be expressed in a simple way in terms of sines and cosines. That is the situation in the case of even dimensions; the case of odd dimensions is different: there the Bessel functions are given as elementary functions. But formula (2) is an extremely important one because it allows us to say a great deal about radial functions and their Fourier transforms because we know a great deal about the theory of Bessel functions. But of course in this kind of an elementary exposition it would take us quite far afield to state explicitly what is known. I shall only be able to tell you what some of the consequences are. Before proceeding with these, there is still a third formula I want to mention.

This is an extremely important formula which in some sense contains both of formulas (1) and (2). It's a formula that is due to Hecke and to Bochner. We see from formula (2) that the Fourier transform of a radial function is a radial function, but a general function in n dimensions has a double dependence. It has a dependence upon how far a point is from the origin and also it has an angular dependence. In order to take into account its angular dependence we have to introduce another basic topic—the notion of spherical harmonics. I don't wish to devote a lot of space to spherical harmonics, but I could not leave this out of even a quick survey of Fourier analysis on Euclidean n space. Consider a spherical harmonic polynomial $P(x)$ of degree k in R^n. What does that mean? It means a polynomial in x, i.e., a polynomial in the variables $x_1, x_2, \ldots, x_n$, which is homogeneous of degree k (the degree of each term is k), and is harmonic. *Harmonic* means that this polynomial is annihilated by the n dimension Laplacian $\partial^2/\partial x_1^2 + \cdots + \partial^2/\partial x_n^2$. Of course the spherical harmonic polynomials of degree 0 are exactly the constants. The spherical harmonics of degree 1 are exactly the linear combinations of the functions $x_1, x_2, \ldots,$ and x_n. So far, the situation is easy. Now the spherical harmonics of degree 2 are a little bit more complicated but one can always write them down. There's a well-known theory of spherical harmonics of any degree in any number of dimensions. The situation is particularly nice when n is equal to 2.* Here of course we are back to classical analysis of one complex variable and harmonic functions of one complex variable. Then everything can be expressed in terms of Fourier series on the circle. (I shall leave that point out altogether.) Suppose I take one of the spherical harmonics of degree k (recall that this is a polynomial) and multiply it by the same function $e^{-\pi|x|^2}$.

The resultant function is essentially its own Fourier transform, except for a small factor which depends on the degree k of the polynomial. We get [13, p. 155]

$$\left(P(x)e^{-\pi|x|^2}\right)^{\wedge} = i^k P(x)e^{-\pi|x|^2}. \tag{3}$$

This is the identity of Hecke. The reason it is important is that the factor $e^{-\pi|x|^2}$ takes care of the radial dependence and is reproduced, and the factor $P(x)$ takes care of the angular dependence. This can be thought of as the basic formula from which the other 2 formulas follow.

4. L^p SPACES AND THE FOURIER TRANSFORM

Now we are going to try to come to grips with our first topic, the Fourier transform. Suppose we have some very good hold on a function f. Say it belongs to a certain very precise class—much more precise in some sense than the Schwartz class $\mathcal{S}$ that I mentioned above. What can we say about the Fourier transform of f? This is of course the basic kind of question that occurs over and over again in Fourier analysis. I shall be able to consider only one infinitesimal part of this very important topic. The main classical theorem along these lines is the so-called Hausdorff-Young theorem. Let me mention it.

---

*In that case, the spherical harmonics of degree k are the linear combinations of $(x_1 + ix_2)^k$ and $(x_1 - ix_2)^k$.

So far we have discussed only one Banach space, the space L^2. As you all know, in Fourier analysis there are other important spaces of this kind. The basic space I want to mention here is the $L^p(R^n)$ space, $1 \leqslant p \leqslant \infty$. This consists of all functions f on R^n which are measurable and for which the p-th norm—$\|f\|_p = \left(\int_{R^n} |f(x)|^p dx\right)^{1/p}$—is finite. That's the situation when p is less than infinity. When p is equal to infinity, we have the space of all functions which are essentially bounded [16, vol I, p. 18]. The Theorem of Hausdorff-Young tells us something about the Fourier transform on the L^p spaces. Again look at the definition of the Fourier transform, $\hat{f}(x) = \int_{R^n} e^{2\pi i x \cdot y} f(y) dy$. As it stands, I cannot put into the integrand a function f in L^p (the integral would not converge) unless the function were in L^1 (in which case the integral defining $\hat{f}$ would converge). So I write down what is called an *a priori* inequality—an inequality not for L^p, but for $\mathcal{S}$ —$\mathcal{S}$ being a dense subclass of L^p (except when $p = \infty$, but I'm not going to consider that case). The Hausdorff-Young theorem, which is a rather remarkable theorem, says that the q-th norm of the Fourier transform is less than or equal to the p-th norm of f. Here $1 \leqslant p \leqslant 2$ and q is the exponent which is conjugate to p in the sense of Holder's inequality, $q = p/(p - 1)$ so that $1/p + 1/q = 1$. In short,

$$\|\hat{f}\|_q \leqslant \|f\|_p, \; 1 \leqslant p \leqslant 2; \quad f \in \mathcal{S}. \tag{4}$$

In particular, on the dense subset $\mathcal{S}$ of L^p, the Fourier transform is a bounded operator with norm less than or equal to 1 from L^p to L^q, and therefore it has a unique bounded extension which maps each L^p function to a function in L^q—a function which of course is defined only almost everywhere. Let us look at the two best special cases of the Hausdorff-Young theorem. The first case is when $p = 2$. Here the conjugate exponent q is also 2 and then (4) is simply a weakening of our earlier statement of Parseval's identity. The other extreme case is when $p = 1$, in which case $q = \infty$. It merely states that if f is an L^1 function, then $\hat{f}$ is bounded, which is obvious by taking absolute values inside the integral in the definition of $\hat{f}$. We have even more than that. The Fourier transform is really a continuous function when f is in L^1; not only is it a continuous function but it's a function which vanishes at infinity, which is the so-called Riemann-Lebesgue lemma [6, p. 7]. For p strictly between 1 and 2, the Hausdorff-Young theorem is usually proved by some sort of convexity argument—an interpolation argument between these two extreme cases [13, pp. 178–183], [7].

5. RESTRICTIONS OF FOURIER TRANSFORMS

I would like now to raise a problem and bring up some more recent results in connection with the Hausdorff-Young theorem. This non-elementary problem arises from the following observation, which I have been told goes back to Laurent Schwartz. Many people have made it independently. Let us now place ourselves in the situation where $n \geqslant 2$. The case $n = 1$ is a very familiar classical situation, but now there are going to be some novel phenomena when $n \geqslant 2$. Let us take a function $f \in L^p(R^n)$ and let us assume, for the moment, that f is radial. This is a familiar way of going about things in n dimensional Fourier analysis. Things are very complicated, but you immediately look for the greatly simplified situation of f being radial, because then the Fourier transform is so neatly given to you by formula (2) and you can try to see what this formula will tell you. Now we examine formula (2) under the assumption that f is in L^p and f is radial.

We make use of the fact that the Bessel function $J_{(n-2)/2}(t)$ is less than a constant multiple of $t^{-1/2}$ as $t \to \infty$. This is a fact of life in the theory of Bessel functions. [13, p. 158]. Of course more precise statements can be made and even used. One uses formula (2) and Holder's inequality—this is the simplest way to get a hold of an L^p situation. One observes the following remarkable thing: the Fourier transform of a radial function is not merely in L^q. Since on the one hand we are taking the Fourier transform of an L^p function with $1 < p < 2$, the Hausdorff-Young theorem tells us that the Fourier transform is some L^q function. An L^q function is a function which is defined only almost everywhere in general. On the other hand $\hat{f}$ is radial. A function which is radial which is defined only almost everywhere is a function of r which is defined only for almost every r. However, it turns out that for certain values of the exponent p, the resulting function is continuous for all $r > 0$. So here is the observation which as I noted is drawn from antiquity (antiquity being 1945 to 1950, I guess).

OBSERVATION: *$\hat{f}(r)$ is continuous for $0 < r < \infty$ when $1 \leqslant p < 2n/(n+1) < 2$.*

The exponent $2n/(n+1)$ I will call the critical L^p exponent in n dimensions. It may look a little strange; it's important nevertheless. So as soon as $p < 2n/(n+1)$, the Fourier transform of a radial function is continuous. Notice that when $n = 1$ the observation is vacuous. When $n = 2$, the critical exponent is $4/3$. This raises the natural problem, can anything be said if the function is no longer radial? Can one make a statement of the same kind as the observation? Let f be a completely arbitrary L^p function. Consider, for simplicity, the behavior of $\hat{f}$ restricted to the unit sphere S^{n-1} of R^n. We have $\hat{f}$ defined on R^n, but defined only almost everywhere on R^n. Is it possible to give a natural definition on the unit sphere so that $\hat{f}$ is a function? The first result of this kind is a rather elementary result, which I observed some time ago; it is, however, quite surprising. It is that even if we no longer assume that the function is radial we can still conclude for certain $p > 1$ that the Fourier transform of an L^p function is well defined on the unit sphere (and hence on any sphere); and it is, in fact, an L^2 function on S^{n-1}. I will therefore define for you a general notion. I will say the (L^p, L^q) restriction property holds in R^n if

$$\left(\int_{S^{n-1}} |\hat{f}(\sigma)|^q d\sigma\right)^{1/q} \leqslant A_{p,q} \|f\|_{L^p(R^n)}$$

holds as an *a priori* inequality for f in the Schwartz class $\mathcal{S}$. If such an *a priori* inequality holds for a given p and q, then I can always say by a limiting argument that the Fourier transform of an L^p function is a well defined L^q function on the unit sphere in R^n. The first result of this kind is a rather elementary one. It shows that there is always some property of this kind, that is an (L^p, L^2) restriction property. In the case of $n = 2$, it says that it holds for all exponents $p < 8/7$. The fraction $8/7$ is a rather ridiculous looking number; what it is in general is $4n/(3n + 1)$. The argument works for any n and shows that if you have a function in $L^p(R^n)$ where $p < 4n/(3n + 1)$, then its Fourier transform's restriction is a well defined L^2 function on the unit sphere. Notice what has happened here. As n tends to infinity, the exponent $4n/(3n + 1)$ tends to $4/3$. For the radial function's restriction property the critical exponent (being $2n/(n + 1)$) tends to 2. The gap between $4n/(3n + 1)$ and $2n/(n + 1)$ is becoming wider, so it's clear that this is probably not the right answer and it will turn out to be not the final answer. This observation was without much interest, I would guess, (except possibly for its shock value) until Charles Fefferman came along and showed it could be used to deal with a basic problem in R^n which is the multiplier problem for the ball. (In effect, what he proved in his Acta paper [2] is that for whatever exponent p we have this (L^p, L^2) restriction phenomenon, for those p we can say some very good things for multipliers which "look like" the characteristic functions of balls.)*So it became interesting to push those results further. In his paper the result is pushed further, and the result goes up to $6/5$ in 2 dimensions, or up to $2(n + 1)/(n + 3)$ in n dimensions. What was in effect proved in that paper was that for $p \leqslant 6/5$ in 2 dimensions the Fourier transform of such a function has a meaning in the L^2 sense on the unit circle. In Fefferman's Acta paper there was also a result which dealt with what happens for any p less than $4/3$. The result was put in final form in a paper of Zygmund as a result in 2 dimensions [17]. Only there in 2 dimensions is the final result known. Here is the 2-dimensional result.

*This paper and a companion piece in the Annals [3] give a very nice overview, and resolution, of the multiplier problem. The reader is urged to at least read their introductions.—*Editor*.

Suppose $p < 4/3$, then the (L^p, L^q) restriction holds with $q = \frac{1}{3}p'$; where p' is the conjugate exponent to p, that is, $q = \frac{1}{3}p/(p - 1)$. In other words, up to 4/3 we always have a restriction, but the L^q class of this restriction gets weaker and weaker as you approach the critical exponent 4/3. But as it passes 6/5 the class L^q is exactly L^2. At 1, it is of course L^∞, because the Fourier transform of an L^1 function is in L^∞. That is presumably the final result of this kind in 2 dimensions. The interesting problem is what holds in higher dimensions? The 2-dimensional situation is somewhat accidental for technical reasons. The basic reason is that in some sense, in every proof of this, or of related phenomena, the exponent 4 plays a crucial role in 2 dimensions. (Note 4 and 4/3 are conjugate exponents.) In higher dimensions the corresponding exponent is less than 4. Now the exponent 4 is a very nice one in Fourier analysis, because a function f being in L^4 is the same as its square being in L^2, and checking if a function is in L^2 can be easily done by looking at the Fourier transform. So for reasons of this kind the exponent 4 is a relatively easy one to deal with and that's why one can push through things in 2 dimensions. In higher dimensions this trick is not available. And so it leaves open an interesting problem. Let me now state the general restriction problem for n dimensions: Prove that you have the (L^p, L^q) restriction phenomenon for $1 \leqslant p < 2n/(n + 1)$ with $q = ap'$. (Recall p' is the exponent conjugate to p.) The number a is not so complicated: $a = (n - 1)/(n + 1)$.*

This turns out to be the anticipated result.† Now, suppose we look at what in principle might be an easier problem: not to do the full (L^p, L^q) restriction problem, but to do merely the (L^p, L^2)

---

*There is a very nice way to see that the right choice of q is not larger than ap'. Let f be the inverse Fourier transform of the characteristic function χ_R of the rectangle R which is formed as follows: Pass two planes perpendicular to the x_1 axis through the points $(1, 0, 0, 0, \dots, 0)$ and $(1 - \epsilon, 0, 0, \dots, 0)$ respectively. Then R is the smallest rectangle with sides parallel to the axes which encloses that portion of S^{n-1} lying between those planes. Here is the picture for $n = 2$ (Figure 1).

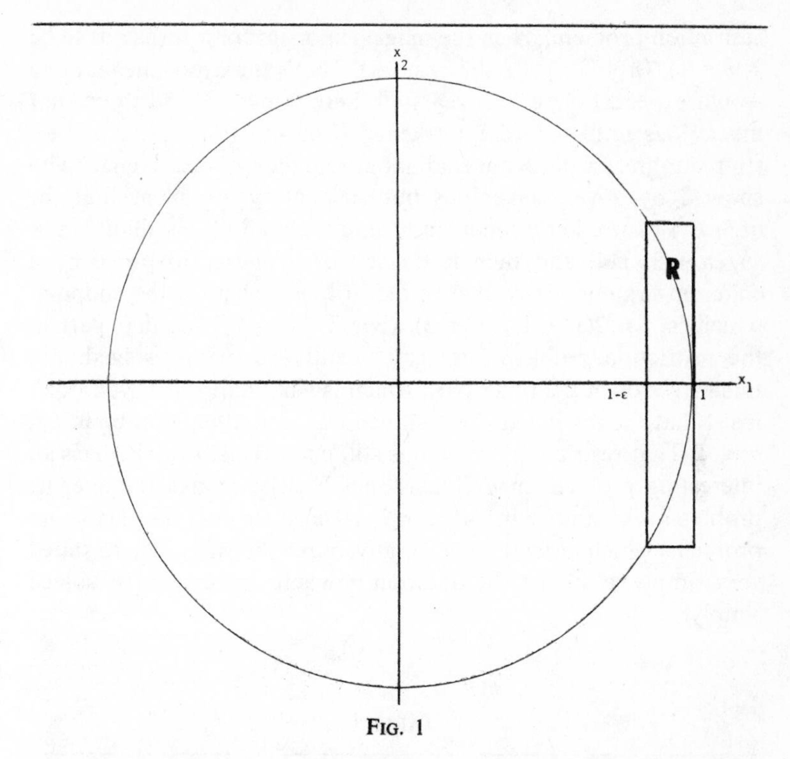

FIG. 1

Since that portion of S^{n-1} lying within R has area on the order of $\epsilon^{(n-1)/2}$ for small ϵ and $\mathcal{F}(f) = \chi_R$, $\|\mathcal{F}(f)\|_{L^q(S^{n-1})} \sim \epsilon^{(n-1)/2q}$. On the other hand, a direct calculation yields

$$\|f\|_p = \left(\int_{-\infty}^{\infty} \left| \frac{\sin \pi x_1 \epsilon}{\pi x_1} \right|^p dx_1 \right)^{1/p} \cdot \prod_{j=2}^{n} \left[\int_{-\infty}^{\infty} \left| \frac{\sin 2\pi x_j \sqrt{2\epsilon - \epsilon^2}}{\pi x_j} \right|^p dx_j \right]^{1/p}$$

$$\sim \epsilon^{(n+1)/2p'}.$$

Letting $\epsilon \to 0$, we see that the hoped for *a priori* inequality can be true only if $(n-1)/2q \geqslant (n+1)/2p'$, i.e., only if $q \leqslant (n-1)/(n+1) \cdot p'$.

†It is easy to see from formula (2) that there can be no restriction problem when $p \geqslant 2n/(n+1)$.

restriction problem. Then the magic exponent for p turns out to be $2(n+1)/(n+3)$ (for n dimensions). That's the exponent that one would expect. For several years all these things seemed to be very mysterious until we were awakened from our dogmatic slumbers (to paraphrase what Kant said about Hume) by Peter Tomas* who showed by a very ingenious but elementary argument that the (L^p, L^2) restriction phenomenon holds for all p less than $2(n+1)/(n+3)$ [14] and then it was an easy matter to prove by a different argument also that in fact it holds even for the endpoint which is $p = 2(n+1)/(n+3)$. (See Table 1.) Thus that part of the restriction problem has very recently been fully solved. But this leaves open the other part, which is what happens if you don't insist that the restricted function be in L^2, but allow it to be in L^q, $q < 2$. That restriction problem is still open. This, I think, gives an interesting problem in n-dimensional Fourier analysis unlike the problems which arise in 1-dimensional Fourier analysis. These are problems which at least have the advantage that they can be stated very simply, although the question is whether they can be solved simply.

TABLE 1

	(L^p, L^2) restriction property holds if p is between 1 and		$(L^p, L^{p'(n-1)/(n+1)})$ restriction property holds if p is between 1 and	f radial implies $\hat{f}$ continuous if p is between 1 and
2 dimensions	$\frac{8}{7}$ early result	$\frac{6}{5}$ best result	$\frac{4}{3}$	$\frac{4}{3}$ (circa 1945 observation)
n dimensions	$\frac{4n}{3n+1}$ early result	$\frac{2(n+1)}{n+3}$ best result	$\frac{2n}{n+1}$ conjecture	$\frac{2n}{n+1}$ (circa 1945 observation)

*At that time a student of A. Knapp.

6. SHARP FORM OF HAUSDORFF–YOUNG THEOREM FOR FOURIER TRANSFORMS

I would like now to pass to another important topic having to do with things related to the Hausdorff-Young theorem. I would like to discuss another rather remarkable theorem. I'm going to ask the following question, that of course wouldn't have been asked had not somebody found an interesting partial answer to it. The question is: Is this Hausdorff-Young inequality, $\|\hat{f}\|_q \leqslant 1 \cdot \|f\|_p$, with the constant 1, the best possible? For a long time people never really thought of raising the question because of the genesis of the subject, which is that Fourier transforms came after Fourier series. If you write down the corresponding Hausdorff-Young inequality for Fourier series, which was the first one that was proved (and the Hausdorff-Young theorem for Fourier integrals was proved as a consequence), that corresponding inequality is obviously best possible with respect to the norm because the function e^{ix} is an extremal function. So the problem was never raised, but about ten to fifteen years ago a Russian mathematician named Babenko proved the remarkable theorem that not only is this inequality not best possible, but he found the best possible constant when p is conjugate to an even integer, i.e., q is an even integer. (We're coming back to 4 – 4, 6, 8, etc.) He found that the best constant is $\left(\dfrac{p^{1/p}}{q^{1/q}}\right)^{n/2}$. Now that looks terribly messy and very meaningless, except that I pointed out to you that the function $e^{-\pi|x|^2}$ is its own Fourier transform. Also it belongs to every L^p class. One easily calculates

$$\|e^{-\pi|x|^2}\|_q = \left(\frac{p^{1/p}}{q^{1/q}}\right)^{n/2} \|e^{-\pi|x|^2}\|_p.$$

Thus the content of Babenko's result is that $e^{-\pi|x|^2}$ is an extremal function (i.e., no function is shrunk less than this one) when p is the conjugate of an even integer. The problem remained open for many years until William Beckner solved it for general p, $1 \leqslant p \leqslant 2$ [1]. The result is the same. Beckner's solution is a very remarkable result, in part because the proof of Babenko cannot be generalized to that case. One needs a different analysis which seemingly begins in the same way, but actually has many different roots.

So it turns out that there is a best constant in the Hausdorff-Young inequality; it is always less than 1 when p is strictly between 1 and 2, and it is the constant that you get merely by testing on $e^{-\pi|x|^2}$ or any of its dilates. (All dilates give the same constant.) That is the Babenko-Beckner theorem. I would like to say something about the proof of this theorem, and then I would like to pass to convolution operators and show how this relates to various positive results and open problems in that area.

So let me give you a vague kind of sketch of what is behind the proof of this theorem. Actually, in order to understand the background of the proof, you have to go back to work that was done by Glimm, Nelson and L. Gross in axiomatic field theory, including certain inequalities involving Hermite functions. I won't delve into this subject, but the Hermite functions have a fundamental connection with the Fourier transform, which is something I will discuss later on. Glimm et al. proved certain inequalities and their point of view was taken over by Beckner. The point is the following: We first of all try to simplify this problem very greatly by writing out an analogue of the Fourier transform and an analogue of the inequality

$$\|\hat{f}\|_q \leqslant \left(\frac{p^{1/p}}{q^{1/q}} \right)^{n/2} \|f\|_p$$

in a very, very simpleminded way. Let $n = 1$. Instead of taking the space of all functions on the real line, we shall only take the two-point analogue of this—we're going to take the space of all functions on a two-point space. What is the two-point space going to be? It will consist of two points symmetrically located about the origin. One is going to be the point $-a$ and the other is going to be the point a. (If you want, you can take them to be -1 and $+1$ and then stretch.) On this two-point space there are only two linearly independent functions. One function is the constant function 1 and the other function is the function x. So the general function is of the form $b + cx$. What will be the Fourier transform on this two-point space? It will leave b alone, but multiply cx by i, $\mathfrak{F} : b + cx \to b + icx$. (The constants b and c are in principle complex numbers.) We have to find the norm on this two-dimensional function space. What I do is take the measure $d\mu(x)$ on the whole real line which assigns a mass of $\frac{1}{2}$ to the point -1 and a mass of $\frac{1}{2}$ to the point 1. Now I consider the following version of the Hausdorff-Young inequality. (Functions will be of the form $c + dx$.) I consider the L^q norm of $\mathfrak{F}(f)$, not with respect to the measure $d\mu(x)$, but with respect to that measure dilated by the factor $\sqrt{q/2}$ —that will be $d\mu(\sqrt{q/2}\,x)$. So the inequality I want is

$$\|\mathfrak{F}(f)\|_{L^q\left(d\mu\left(\sqrt{q/2}\,x\right)\right)} \leqslant 1 \cdot \|f\|_{L^p\left(d\mu\left(\sqrt{p/2}\,x\right)\right)}.$$

But now I have the norm 1. What we've done by putting in these factors $\sqrt{q/2}$ and $\sqrt{p/2}$, is to transform our problem into a problem about an operator on a very simple L^p space to another equally simple L^q space having norm less than or equal to 1–a contraction operator. This is the two point result—the analogue of the Babenko–Beckner theorem. This has to be proved and, as you can guess, is an elementary looking inequality which involves two parameters, i.e., really one parameter after a change of scale. Once that inequality has been proved, what you do is the following: Start with this two-point space with the probability measure I introduced and take the product of this space with itself any number of times, taking the induced transformation which is still a contraction when you take the product of these transformations on this larger space, and then pass to the limit in more or less the same spirit that you take the product of two-point spaces with probability measures on them. When you pass to the limit, lo and behold!, what do you get? You get the Gaussian distribution. Now if you do this process correctly (and this process is actually related to the theory of Hermite functions), you will then come out with the desired result, which leads to the best constant, $\left(\frac{p^{1/p}}{q^{1/q}}\right)^{1/2}$. To extend to n dimensions is easy. Everything is multiplicative so the result follows with $\frac{1}{2}$ replaced by $n/2$.

I would like to say just one word more about the background of this theorem which I think seems still quite mysterious. I will venture to predict that the ideas in it will turn out to be very important insofar as they give us a new kind of insight into the Fourier transform. I also would like to point out that as far as I'm concerned, I really don't understand what is going on and why it really works. The thing it is related to is of course the notion of Hermite functions so I will say something about the Hermite functions and their role in the Fourier transform.

In order to keep things elementary, let's stick to the case of $n = 1$. The Fourier transform maps the function $e^{-\pi|x|^2}$ into itself —$\mathfrak{F} : e^{-\pi|x|^2} \to e^{-\pi|x|^2}$. It also has another generalization, not in the direction of Hecke's identity (equation (3)), but in the direction of Hermite functions which was the starting point for one of the classical proofs of the Plancherel theorem which can be found in Wiener's book [15, pp. 46–57]. That is, if you take suitably defined Hermite polynomials (that is, not the usual normalization because the usual normalization has $e^{-\frac{1}{2}|x|^2}$ and we need a π in the exponent) $H_n(x)$ of degree n, and you take the Fourier transform of the product $H_n(x)e^{-\pi|x|^2}$, then this is mapped into itself, except for the factor of i^n—

$$\mathfrak{F} : H_n(x)e^{-\pi|x|^2} \to i^n H_n(x)e^{-\pi|x|^2}. \tag{5}$$

How do we get the polynomials $H_n(x)$? Well, we take a certain measure of total measure one on the real line, that's the measure $d\mu(x)$ which will play, in some sense, the role of the two-point mass that I considered earlier, and that measure is $d\mu(x) = 2^{\frac{1}{2}} e^{-2\pi|x|^2}\, dx$. Suppose you take this measure, and with respect to this measure you orthogonalize the ordinary polynomials; then you'll get the normalized Hermite polynomials with this normalization. The action of the Fourier transform is given by (5), and that of course allows you to induce another transformation which can be called K, which is merely the mapping of $H_n(x)$ into $i^n H_n(x)$—

$$K : H_n(x) \rightarrow i^n H_n(x).$$

You can use K on linear combinations of the $H_n(x)$. It becomes a unitary mapping of $L^2(d\mu)$ to itself. That's really now a translation of the unitary property of the Fourier transform. Finally, what the sharp, the best constraint inequality for the Fourier transform is, is an inequality about the mapping K—saying that with respect to certain norms, K has operator norm less than 1. The situation is this

$$\|K(f)\|_{L^q\left(d\mu\left(\sqrt{q/2}\,x\right)\right)} \leqslant 1 \cdot \|f\|_{L^p\left(d\mu\left(\sqrt{p/2}\,x\right)\right)}, \qquad 1 \leqslant p \leqslant 2, \tag{6}$$

where again the norms are taken not with respect to $d\mu(x)$, but with respect to appropriate dilations of $d\mu$. If you take (6) and unscramble it by the appropriate changes of variables, then you will get the Babenko-Beckner theorem for $n = 1$. It is this result that, in some sense, can be obtained by applying a limiting process to the procedure of taking those two-point inequalities. What is involved is passing to infinite products and then taking a certain subspace of the resulting space of functions—this will give the inequality (6). This completes what I mean to say on my first topic —the Fourier transform.

7. CONVOLUTION OPERATORS

Now I would like to turn to convolution operators and mention a variety of problems which arise partly in view of the results that have been discussed above. The basic inequality about convolution operators is the inequality of Young. Of course, historically speaking the importance of the inequality of Young for convolution operators was and is intimately tied up with the proof of the Hausdorff-Young theorem. I'll now display for you the inequality of Young and raise a variety of questions. What one does is take the convolution of two functions, $f * g = \int_{R^n} f(x-y) g(y)\, dy$. Never mind, for the moment, in what sense the integral exists. Of course, the basic property about this convolution is that when it and the Fourier transforms of f and g all make sense, the Fourier transform of this convolution is the product of the Fourier transforms—$\widehat{(f*g)}(y) = \hat{f}(y)\hat{g}(y)$. Now Young's inequality deals with the question of finding the size of the convolution $f*g$ in terms of the sizes of these functions f and g. Again we're going to be dealing with L^p's in stating the theorem. Young's inequality for convolutions is the following

$$\|f*g\|_{L^r} \leqslant 1 \cdot \|f\|_{L^p} \cdot \|g\|_{L^q}, \qquad \frac{1}{r} = \frac{1}{p} + \frac{1}{q} - 1. \tag{7}$$

Of course all the exponents p, q, and r are assumed to lie between 1 and ∞. That puts certain restrictions on formula (7).

It is possible to write down a similar Young's inequality, of course, in the case of Fourier series. And this Young's inequality for Fourier series is, in some sense, best possible. This is for the same reason that the Hausdorff-Young theorem for Fourier series was best possible, namely, that there can be no improvement of the Fourier series version of inequality (7) if f and g are to range over all L^p and all L^q functions, and that you can see merely by plugging in $f = g = 1$. However, inequality (7), which is one of the most basic inequalities which is discussed in every textbook, with this constant 1 is not sharp on R^n. It can be seen that there is a better constant than the constant 1. The best constant can be found as a result of the Babenko-Beckner theorem in already a substantial special case. The substantial special case corresponds to the situation when $r \geqslant 2$, but $p \leqslant 2$ and $q \leqslant 2$. Let me give you an example of how one goes about finding the best possible inequality in that case. A simple example arises in this situation—

when $r = 2, p = 4/3$ and $q = 4/3$. Here the exponent $4/3$ is again rearing its ugly head. As we have already mentioned, the importance of the exponent $4/3$ is that its conjugate exponent is 4. How do we obtain a sharper inequality? If we know that f is in $L^{4/3}$, then we have control of its L^p norm. Let's say that its L^p norm ($= L^{4/3}$ norm) is 1 and let's say that the L^q norm ($= L^{4/3}$ norm) of g is 1. Then you take the Fourier transform of both sides of inequality (7). Now the Fourier transform of an $L^{4/3}$ function is an L^4 function by the Hausdorff-Young theorem; but by the sharp form of the Hausdorff-Young theorem the L^4 norm is better than 1—it's the value you get from the exponential function $e^{-\pi|x|^2}$. So the Fourier transform of f is an L^4 function and the Fourier transform of g is an L^4 function. The product, therefore, is an L^2 function (by Holder's inequality) and the L^2 norm of that is better than what you would get from inequality (7). You can actually get the best possible result. Because if you take dilations of the exponential $e^{-\pi|x|^2}$, they are really closed under convolution, as you can see by taking the Fourier transform. So you get a certain constant by just taking exponentials of the form $e^{-a|x|^2}$. That would be a lower bound for the best Young's inequality constant, but actually the true bound is no bigger than this in this case, as is seen by the use of the sharp form of the Hausdorff-Young theorem. So the problem that arises is to prove the corresponding result for all p's q's and r's satisfying $1/r = (1/p) + (1/q) - 1$, subject only to the restriction that p, q, and r are between 1 and ∞. This seems to be a rather basic problem because we are dealing here with positive functions. To solve this problem, is to solve it merely for positive functions where there is no cancellation. We'd like to have a better hold of how convolutions behave with respect to L^p norms. So, the first problem I will raise on convolutions is

PROBLEM: Prove a sharp form of Young's inequality.*

---

*The problem has been resolved recently [1].

I earlier mentioned something to the effect that this particular problem has no analogue for Fourier series, because there we know that if you range over all functions f and g, the constant can be no better than 1. But I would like now to point out two facts to you which reveal that there is also a very interesting problem in the periodic case. I shall now say a little bit about Fourier series, even though the title of my article is "Harmonic analysis in R^n." I shall make two observations which will lead to a problem. Our first observation deals with the Poisson kernel. It doesn't really need to be the Poisson kernel; it could be almost anything else like it. But it suffices to say that we're going to deal with the Poisson kernel.

Now we're taking functions defined on $[0, 2\pi)$. We're going to take the usual $L^p(0, 2\pi)$ spaces, and we're going to take the familar Poisson kernel, $P_r(\theta)$, which I shall write down in the following form:

$$P_r(\theta) = \frac{1}{2\pi} \cdot \frac{1 - r^2}{1 - 2r \cos \theta + r^2}.$$

This is not the form that is usually used (*editor's remark*: $P_r(\theta)$ is classically the above kernel multiplied by π [16, vol. I, p. 96]), but I want to write it down in such a way that its total integral is equal to 1, $\int_0^{2\pi} P_r(\theta) d\theta = 1$. I'm going to take the mapping $f \to P_r * f$ and make the following claim: Suppose p and q are any pair of exponents, $p < q < \infty$; then of course for each $r < 1$, the mapping $f \to P_r * f$ is bounded from L^p to L^q. The meaning of this operator has a very simple but basic interpretation. We start with a function f which is to be thought of as being defined on the boundary of the unit circle (identifying θ of $[0, 2\pi)$ with $e^{i\theta}$ of the complex numbers) and pick an $r < 1$ and consider the concentric circle of radius r (lying within the unit circle) and consider the corresponding harmonic function which has the boundary values f (the solution to Dirichlet's problem for the unit disc). Finally we take the restriction of that harmonic function to the circle of radius r—this restriction is exactly the convolution $P_r * f$. Thus we are saying that if we start with any L^p function defined on the unit circle, the restriction of its harmonic extension to any interior circle of radius r will be a function in L^q, for any fixed r. Of course that is an easy thing to prove; it's a simple estimate if you wish. It's a simple application of the analogue of Young's inequality here because the Poisson kernel belongs to all the L^q spaces. But now what I claim to you is that fixing p and q, there exists an $r = r(p, q)$ which depends on p and q and which is greater than 0, so that not only is $f \to P_r * f$ a bounded operator from L^p to L^q, but it has norm less than or equal to 1—again, it's a contraction operator. (Note that this will then hold for all $r \leqslant r(p, q)$ since $\|P_r * f\|_q \nearrow 1$ as $r \nearrow 1$.) This is a little bit remarkable because, of course, 1 is the least constant that you could get. (Let $f = 1$.) Young's inequality would give $\|P_r * f\|_q \leqslant \|P_r\|_\alpha \|f\|_p \leqslant \|f\|_p$ provided $\alpha = 1/[1 - (p^{-1} - q^{-1})] = 1$; but of course $p < q$ and so the L^1 norm is not what you can use. Nevertheless, it's possible to prove the claim. It's not too difficult to prove it. It really comes out of the kind of reasoning that was used by the people in quantum field theory—Glimm, etc. I think it's somewhat surprising from the point of view of classical Fourier series. (For example, $\|P_r * f\|_q \nearrow \|f\|_q$ and $\|f\|_q > \|f\|_p$, in general, so $r(p, q)$ cannot be *too* close to 1.) So there is something else besides the fact that the function g belongs to some L^q space which controls the convolution inequality (7). Because here $\|P_r\|_\alpha$ is greater than 1 so one does not expect the result from considering the sheer size of P_r but still you get the inequality, $\|f * P_r\|_q \leqslant \|f\|_p$. So part of the question is to understand this better. One other question is to know what is this $r(p, q)$ so that as soon as we are inside that r, the norm is less than or equal to 1? Let me point out one more thing, and then I can raise the general problem.

If one looks carefully in Zygmund's book, *Trigonometric Series*, [16, vol. 11, p. 146 and p. 333, lines 11–14], one will find a rather remarkable example which has a long background which is described. The example is the following: There exists a finite non-negative measure $d\mu$ on the circle that is the Fourier-Stieltjes series of a monotonic function. This measure is totally singular, that is, it lives on a set of Lebesgue measure zero. It has the following property: If we write down its Fourier-Stieltjes series,

$$d\mu \sim \sum a_n e^{in\theta}$$

the coefficients $\{a_n\}$ decrease as rapidly as they are allowed to under the circumstances; let's say, $|a_n| \leq c/\sqrt{n}$. There is this example in Zygmund's book. If one looks at another theorem in Zygmund's book, [16, vol. II, p. 127, with $\alpha = \frac{1}{2}$], one then realizes that the following thing is true: Suppose I take the operator of convolving with this singular measure $- f \to f * d\mu$. This is a positivity preserving operator because $d\mu$ is presumed to be nonnegative. This measure $d\mu$ can in no sense be thought of as belonging to any of the L^p spaces, where $p > 1$. It could be thought of as belonging, in some sense, to the limiting situation of L^1; but not to any L^p spaces because of course it is a singular measure. But this convolution operator $f \to f * d\mu$ leads to a bounded operator from L^p to L^q which is better than one would expect again from the situation of Young's inequality. The result is that there is a definite improvement in q for certain p. We have $1/q = (1/p) - \frac{1}{2}$, so there is this definite improvement of $\frac{1}{2}$, as long as $1 < p \leq 2$, and $q \geq 2$. That's one thing that can be said immediately by an examination of two parts of the theory.

Now we are faced with the general problem. Consider convolution operators with positive kernels. What is meant by a positive kernel? It means that you're dealing with a convolution operator—that's an operator which commutes with translations and preserves non-negative functions [13, p. 26]. These operators don't involve any cancellation. We would like to know, what is the condition on the function g, or on the kernel g, or on the measure g, so that $f \to f * g$ will be a bounded operator from L^p to L^r? Of course, that g is in L^q (where $1/r = (1/p) + (1/q) - 1$) is a sufficient condition by Young's inequality. But it's very far from necessary as these kinds of examples show. So here's a very basic problem, I think. Convolution operators are a very basic kind of operator in Fourier analysis. However, these are convolution operators which allow no cancellation and it is asked to characterize (if possible, in terms of the size of the measure $d\mu$, whatever that means) the condition of $f \to f * d\mu$ yielding a bounded operator from an L^p space to an L^r space.

8. SYMMETRY CONSIDERATIONS

I'd like to say something else about convolution operators. For this I will now return to considerations of symmetry. When you impose considerations of symmetry you are restricted to very few kinds of convolution operators. The primary convolution operators that arise are, of course, the identity operator—a trivial kind of convolution operator with the measure $d\mu$ being the Dirac measure—or the Hilbert transform and its generalizations. I'd like to say a few words about this, and then a few words about the Hilbert transform.

We now regress to some very elementary considerations. Let us for the moment restrict our considerations back to the line R^1 and to the very basic space $L^2(R^1)$. Let us ask for a transformation T which is a bounded operator on $L^2(R^1)$ and which satisfies the maximal amount of symmetry. Now remember the symmetries we have discussed in R^n are dilations, translations and rotations. So we shall assume

1. T commutes with translations;
2. T commutes with dilations; and
3. T commutes with $x \to -x$, that is with the mapping that sends $f(x)$ to $f(-x)$.

(I was a little ambiguous by what I meant by rotations; I mean proper and improper rotations. The only proper rotation on the real line is the identity. The improper rotation is the mapping from x to $-x$.) There is only one operator which satisfies all three of these properties. As you well know, that's the constant multiple of the identity, $T = cI$ [11, p. 38]. So much for the characterization of the identity transformation.

One of the reasons the Hilbert transform is so interesting is that it has a similar characterization. We ask now a similar question by merely modifying the third property, and this leads us to the Hilbert transform on the real line. Well this is one way that someone might be led to considering the Hilbert transform. There are lots of other ways, of course, and they're all related. So now we ask for T to be a bounded operator on $L^2(R^1)$ which

1. commutes with translations,
2. commutes with dilations, and
3. anti-commutes with $x \to -x$.

Conditions 1 and 2 are as before, but condition 3 is different—it says that T is an odd mapping. And, lo and behold, the fact is that then T is a constant multiple of the Hilbert transform, $T = cH$. Of course I haven't told you what the Hilbert transform is. I shall now give you a formal definition. The usual way of doing this is to have a certain factor $1/\pi$, so that

$$Hf(x) = \frac{1}{\pi}\int_{\infty}^{\infty} f(x-y)\frac{dy}{y} = \frac{1}{\pi}\lim_{\epsilon \to 0}\int_{|y|>\epsilon} f(x-y)\frac{dy}{y}. \tag{8}$$

This is a convolution integral, but it's a convolution of a non-absolutely convergent type—$\int |f(x-y)| dy/|y|$ is in general divergent since $\int dy/|y|$ doesn't converge. This is why the integral has to be taken in the principal value sense. Even the definition of the transformation given here is a non-elementary one, because the definition says that for appropriate functions f, the limit of $\int_{|y|>\epsilon} f(x-y)dy/y$ (as $\epsilon \to 0$) is well defined almost everywhere (and it then has certain properties). But that gets you into the theory of the Hilbert transform, and I don't really want to deal with the theory of the Hilbert transform but only to mention the Hilbert transform. The point is that the Hilbert transform can be defined this way, and it does lead to a very basic operator on L^2. It leads to a unitary operator on $L^2(R^1)$. But again, it's defined not only on the L^2 space, but on the L^p spaces. It leads to a bounded operator on L^p if p is strictly between 1 and ∞ [13, pp. 186–188]. So that is the Hilbert transform.

Once one knows the Hilbert transform and its many properties on the real line, one can begin to generalize the situation to R^n. I can say almost nothing about the possibility of generalizing this operator and studying properties of the generalizations in the setting of R^n [11]. I will merely assert that there is a very rich theory of things like the Hilbert transform in R^n. One way to find what some of these generalized operators in R^n ought to be is by asking the same sorts of questions that we asked here, but now in the appropriate language for R^n. If you ask those questions in the right way, you are led to a variety of generalizations of the Hilbert transform and various other kinds of singular integrals which play a basic role in the n-dimensional theory at present.*

9. NON-ISOTROPIC DILATIONS

I would like to pursue, however, for one moment, a property of the Hilbert transform. Let us look more closely at one of the symmetry properties I discussed earlier, but now in R^n. This is the property associated with dilations. I will devote a little bit of time to discussing dilations in R^n and their generalizations and what you can do if you play with this notion.

Let me remind you what I call dilations. A dilation was a mapping which maps the point x into $\delta \cdot x$ where δ was positive, $x \to \delta \cdot x$, or in coordinates,

$$(x_1, x_2, \ldots, x_n) \to (\delta x_1, \delta x_2, \ldots, \delta x_n).$$

So this is a dilation which dilates the same amount in every direction. These are the standard kinds of dilations we deal with. To be more precise, you can call them the *isotropic dilations*. As I stated (and hinted), much of standard Fourier analysis on R^n is built with these dilations or has a certain invariance property with

---

*It is here that the theory of H^p spaces enters in a decisive way.

respect to dilations. Or you may look at a theorem and see how it transforms under these dilations. This of course comes up over and over again. Let me just remind you that when we dealt with the basic identity of Hecke (equation (3)) where I wrote $P(x)e^{-\pi|x|^2}$, P was a homogeneous polynomial of a given degree and homogeneity. This is a statement having to do with these dilations. And of course we saw that the Hilbert transform is a basic operator on the real line which has an invariance property with respect to these dilations on the real line.

For certain other problems that arise, it is desirable to consider other kinds of dilations which one might call non-isotropic dilations. One is led to this, not merely because one wishes to do things for the sake of generalizing every concept at every possible stage, but because one is led to this by a variety of problems. What is a typical kind of non-isotropic dilation that one could consider? One could consider exponents $a_1, a_2, \ldots, a_n$ that are positive numbers (strictly > 0), but not all the same. Now one could consider a new dilation

$$\delta \circ x = (\delta^{a_1}x_1, \delta^{a_2}x_2, \ldots, \delta^{a_n}x_n).$$

One still has a one-parameter multiplicative family of dilations, when $\delta > 0$, but these dilations dilate differently in different coordinates. There is a theory of generalizations of the Hilbert transform due to B. F. Jones, G. Fabes, N. M. Riviere, Lizorkin, etc., which is built on these dilations in analogy with the isotropic dilations.

A very simple problem can be raised now if you consider non-isotropic dilations. You can look and see what non-isotropic dilations do to any point. Let's look for a moment at two-dimensional space and let's look at a point x different from the origin, and let's subject the point to the family of isotropic dilations. Clearly what is happening then is that if you consider all isotropic dilations of this point you get the ray passing through the origin and x and going off to infinity (Figure 2(a)). However, if I were to consider a non-isotropic family of dilations, I would get a certain curve passing through the origin and x, going out to infinity. A very natural problem can be raised. The kind of curve that we get here (let me parametrize it by t) is of the following form:

$$\gamma(t) = (A_1(\operatorname{sgn} t)|t|^{a_1}, A_2(\operatorname{sgn} t)|t|^{a_2}, \ldots, A_n(\operatorname{sgn} t)|t|^{a_n}).$$

The curve in the isotropic case was only defined for positive t. Therefore I am free to extend to negative t as I choose. Thus, I put in the factors sgn t to make the curve odd. This is a typical curve, $\gamma(t)$, which is an invariant object with respect to the class $x \to \delta \circ x$ of non-isotropic dilations. For example, if we let $n = 2$, $a_1 = 1$, and $a_2 = 3$, then choosing the constants A_1 and A_2 to make the curve pass through x, we have the situation depicted in Figure 2(b).

You might ask, is there a meaningful notion of the Hilbert transform associated to the curve $\gamma(t)$, and therefore to these dilations—$x \to \delta \circ x$. In view of the definition given above (equation (8)), the Hilbert transform should be the following transformation:

$$H^{\gamma}(f)(x) = \frac{1}{\pi} \int_{-\infty}^{\infty} f(x - \gamma(t)) \frac{dt}{t}.$$

This is the Hilbert transform that depends on the curve γ. It's a mapping, in principle, on functions belonging to $L^p(R^n)$. But, of course, this definition—it is the analogue of the one given in equation (8)—must be given some sort of meaning in terms of principal value limits. The question is, does in general this Hilbert transform make any sense? Is it a bounded operator? Does it have some of the same properties as the classical Hilbert transform on the real line? This might be said to be one of the basic building blocks for the Fourier analysis on R^n, if you look at Fourier analysis from the point of view of these non-isotropic dilations instead of the isotropic dilations. So that's the question that can be asked.

In the special case when all of the a_i's are equal to 1—that's the isotropic dilation—what this becomes, basically, is the ordinary Hilbert transform, but taken on a straight line through the origin.

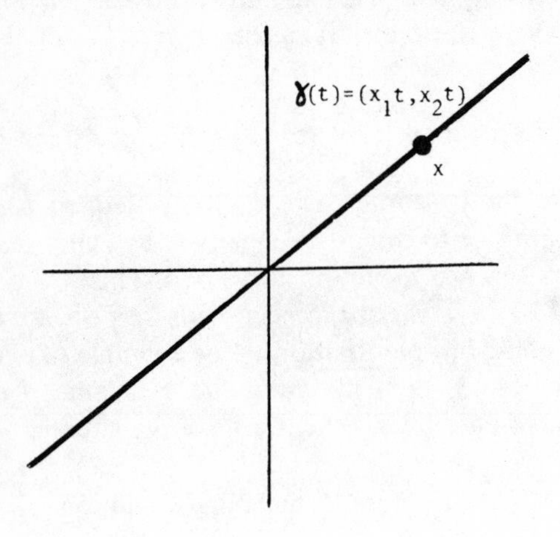

isotropic case

(a)

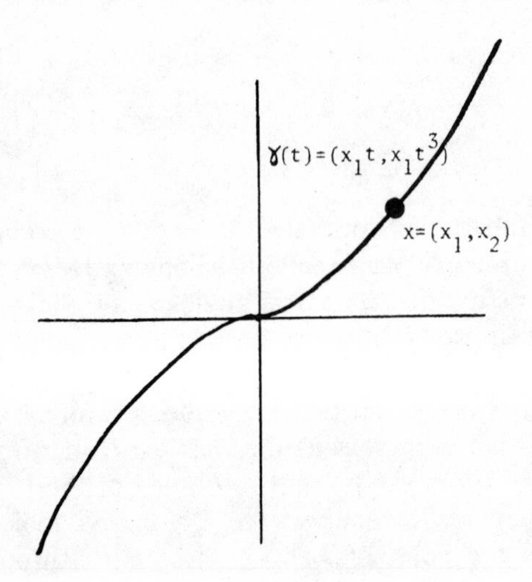

non-isotropic case

(b)

FIG. 2

And the properties of this "Hilbert transform", taken on a straight line through the origin, are then very simple consequences of the one-dimensional theory. What one does, is one breaks up the whole space into lines which are parallel to that straight line. One uses the one-dimensional theory associated with the ordinary Hilbert transform (equation (8)) and then one integrates, so to speak, on all of the straight lines. Then one obtains the fact that this Hilbert transform, when we are dealing with the case of a straight line, is a bounded operator from $L^p(R^n)$ to itself, for p between 1 and ∞—which is merely, as I said, a consequence of the 1-dimensional result.

Now the question is, is there a similar theorem in the case of these non-isotropic dilations? The answer is yes. In a recent result, which was proved by Nagel, Riviere, and Wainger, it was shown that in this generality (that is, for the mapping H^γ defined above) the corresponding Hilbert transform along curves is again a bounded operator on L^p. What is remarkable about this theorem is that its proof is not a generalization of the proof of the straight line case. That is, one uses some extra properties of the curve $\gamma(t)$. The curve gives you some extra smoothness property on the Fourier transform that a straight line doesn't give you. So there is a different argument than in the situation of the straight line. This leads one to a very general philosophic principle of Fourier analysis in R^n which I'd like to state. That is that curves—curved lines, curved surfaces, etc.,—are more interesting and richer than straight lines or linear varieties. We have seen two examples of this kind of phenomenon—not explicitly, but I've alluded to two kinds of examples of this basic type of phenomenon in these lectures. The first example was the restriction problem. We saw that in dimension greater than or equal to 2, if you took an L^p function and you took its Fourier transform, then for certain $p > 1$, there is a meaning to restricting the Fourier transform to a sphere. Such restriction properties also hold for other curved varieties. But clearly there can be no meaning to restricting a function to a straight line; because that immediately reduces the problem to a corresponding problem on the real line and the corresponding statement that if you have an L^p function where $p > 1$, the

Fourier transform of it cannot be restricted to a single point. So, in this case, the situation is that curved varieties are in some sense richer than straight varieties. The proof of the Nagel-Riviere-Wainger theorem (which we cannot give here) is another example of this phenomenon [9].

10. OTHER "TRANSLATIONS" ON R^n*

I have said something now about Fourier transforms. I have said something about convolution operators. And I have said something about symmetry properties of R^n. Now R^n was our underlying space and our three symmetry properties were what? Well, they were translations, they were dilations, and they were rotations. The notion of dilation we have modified already, and this has led to some interesting different kinds of problems. The notion of rotation of R^n can also be modified. This is a very interesting subject in itself of which I will say almost nothing. The reason rotations play such a basic role here is because in the definition of the Fourier transform we took the inner product $x \cdot y$ to be a particular bilinear form (which is $x_1y_1 + x_2y_2 + \cdots + x_ny_n$). The group which preserves that inner product is the group of rotations. Of course, by a mere simple change of variables one could change this form $x_1y_1 + \cdots + x_ny_n$ to another one without really harming the fact that you have a good Fourier transform. For example, write the form as $x_1y_1 - x_2y_2 - x_3y_3 - \cdots - x_ny_n$—that doesn't really change much. But then it's clear that what you want to do is consider not the ordinary rotations but the Lorentzian rotations—the rotations which correspond to indefinite quadratic forms.† One can play this game in a more and more complicated way and one can try to study the Fourier transform from the point of view of certain linear transformations on the underlying R^n space. You study the invariance properties which derive from various kinds of linear transforma-

*For further background on these matters see my article [12].

†The subject of the Fourier transform from the point of view of indefinite quadratic forms has been treated systematically by R. Strichartz.

tions. That's a subject which brings one closer to the theory of harmonic analysis and group representations [10]. So I will pass over the subject of modifying the rotations.

I will now come to grips with the question of modifying the translations. It may seem to you that we are now pursuing generalization for the sake of generalization, but I want to assure you that this is not quite the case. I'm going to look at R^n—the Euclidean space of n dimensions—but with a different translation structure. (Then I shall explain to you why one wants to use this different translation structure.) The simplest example arises when I consider the situation of R^{2n+1}—I shall now, for the moment, consider odd dimensional Euclidean spaces. (Of course, later on, all this can be generalized.) It will be convenient to gather the first $2n$ real variables into n complex variables, and to leave the last real variable alone. So I shall write

$$R^{2n+1} = C^n \times R^1 = \{(z, t)\},$$

where $z = (z_1, z_2, \dots, z_n)$ belongs to C^n and t is a real number. But now, instead of taking the usual multiplication law (of vector addition), I will take the multiplication law to be addition with one modification which is very important for this context. Let me write down for you the multiplication law that I'm going to take. We have

$$(z, t) \cdot (z', t') = \left(z + z', t + t' + 2\,\mathrm{Im}(z \cdot \overline{z'})\right), \tag{9}$$

where $z \cdot \overline{z'} = z_1\overline{z'_1} + z_2\overline{z'_2} + \cdots + z_n\overline{z'_n}$—the complex inner product—is a complex number, so that the definition makes sense. Notice that this is just like the standard addition in the first n complex variables. It can be checked that this multiplication law makes the underlying Euclidean space R^{2n+1} into a group, but the addition law is no longer commutative. (E.g., $(i, 0) \cdot (1, 0) = (1 + i, 2) \neq (1, 0) \cdot (i, 0)$.) It is one step (because of the quadratic term $\mathrm{Im}\, z\bar{z}'$) removed from commutativity. This is a well-known group nowadays. It's not the Euclidean group of R^{2n+1}, but it's the so-called Heisenberg group. I don't want to get into the theory of the Heisenberg group. The only point that I want to make in this article is that the Heisenberg group and other groups of this same kind of character are very important for lots of questions which relate to harmonic analysis. First of all, a trivial remark. The dilations that I would like to assign to the Heisenberg group cannot be the isotropic dilations, but must by necessity be the non-isotropic dilations. If the dilations are going to be such as to be compatible with the multiplication property (9), then if I dilate $z \to \delta z$, I must dilate $t \to \delta^2 t$. Because then

$$\begin{aligned}[\delta \circ (z,t)]\cdot[\delta \circ (z',t')] &= (\delta z, \delta^2 t)\cdot(\delta z', \delta^2 t') \\ &= (\delta z + \delta z', \delta^2 t + \delta^2 t' + \delta^2 \operatorname{Im} z\overline{z'}) \\ &= \delta \circ [(z,t)\cdot(z',t')],\end{aligned}$$

which is the analogue of $\delta \circ (x+y) = \delta \circ x + \delta \circ y$ in the Euclidean case. So if I consider the Euclidean space of dimension $2n+1$ and give it this group structure, then I must naturally be led to the non-isotropic dilations.

Finally, you might ask, what is the interest of the Heisenberg group, and why does it occur at all? Now comes a brief explanation of why, if you follow this thing through, you might be led to the Heisenberg group. First let us go back to the standard Fourier analysis on the real line—R^1. One of the basic properties of this Fourier analysis which I have had no chance to explain, but which must be explained in any introduction, is that the Fourier analysis of R^1 is intimately connected with the theory of holomorphic functions in the upper half plane and the theory of harmonic functions in the upper half plane [4].

This is one of the places where the Hilbert transform comes in in a basic way. So the basic operator, which is the Hilbert transform, is a convolution operator on the real line with the usual additive structure of the real line, and that is intimately connected with holomorphic functions of the upper half plane. Of course holomorphic functions in the upper half plane are intimately connected to holomorphic functions in the unit disc. The two theories are very closely related (the Cayley transform $w = (z -$

$i)/(z + i)$ maps the upper half plane onto the unit disc)—I will not say any more about it. Let us suppose we are looking for an n-dimensional generalization, in the sense of n complex variables, of the unit disc or of the upper half plane. What presents itself? Well, there's a very natural generalization of the unit disc to several complex variables. In that case what one is led to is the unit ball in complex n space. Now you ask, knowing the unit ball in complex n space, what is the analogous upper half plane? One can write down a similar kind of Cayley transform which leads one to an analogous upper half plane, except its boundary is not quite flat, which it cannot be. Then when one looks at the boundary of this "upper half plane" it is identifiable with a Euclidean space, but its natural structure of translations is not the Euclidean structure, but is exactly the structure of the Heisenberg group. Therefore the Fourier analysis on the Heisenberg group is naturally connected with the problems which grow out of several complex variables in the same way as the classical Fourier analysis may be said to have grown out of considerations of holomorphic functions in the upper half plane. One is led to a variety of convolution operators and problems which reflect the Heisenberg group. Other problems lead one to other structures on Euclidean n space which have a group structure which is not exactly the same as translations and which also has built in together with the structure the dilations. These occur in the work of Knapp and others [8] in the theory of intertwining operators and they also occur in a variety of problems involving differential equations. Incidentally, the differential operator which for the Heisenberg group plays the same role as the Laplacian for (the usual additive structure of) R^n is

$$\sum_{j=1}^{n}\left\{\frac{\partial^2}{\partial z_j \partial \bar{z}_j} + |z_j|^2 \frac{\partial}{\partial t^2} - i \frac{\partial}{\partial t}\left(z_j \frac{\partial}{\partial z_j} - \bar{z}_j \frac{\partial}{\partial \bar{z}_j}\right)\right\},$$

with

$$\frac{\partial}{\partial z_j} = \frac{1}{2}\left(\frac{\partial}{\partial x_j} - i \frac{\partial}{\partial y_j}\right) \quad \text{and} \quad \frac{\partial}{\partial \bar{z}_j} = \frac{1}{2}\left(\frac{\partial}{\partial x_j} + i \frac{\partial}{\partial y_j}\right).$$

This operator occurs in the theory of the Cauchy-Riemann equations in several complex variables, and is an example of an important class of so-called "hypo-elliptic" differential operators.* I have just mentioned these last few advanced topics, for which I hope I will be forgiven, to give you at least an impression that the classical subject of analysis in R^n is far from a closed subject. There are many further avenues for research.

REFERENCES

1. Beckner, W., "Inequalities in Fourier Analysis on R^n," *Proc. Nat. Acad. Sci.*, **72** (1975), 638–641.
2. Fefferman, C., "Inequalities for strongly singular convolution operators," *Acta Math.*, **124** (1970), 9–36. MR **41** #2468.
3. ——, "The multiplier problem for the ball," *Ann. of Math.*,(2) **94** (1971), 330–336. MR **45** #5661.
4. ——, "Harmonic analysis and H^p spaces," this volume.
5. Folland, G., and E. M. Stein, "Estimates for the $\bar{\partial}_b$ complex and analysis on the Heisenberg group," *Comm. Pure Appl. Math.*, (to appear).
6. Goldberg, R. R., *Fourier Transforms*, Cambridge Tracts in Math. and Math. Physics, No. 52, Cambridge University Press, Cambridge, 1965.
7. Hunt, R., "Developments related to the a.e. convergence of Fourier series," this volume.
8. Knapp, A., and E. M. Stein, "Intertwining operators for semisimple groups," *Ann. of Math.*, **93** (1971), 489–578.
9. Nagel, A., N. Riviere, and S. Wainger, "On Hilbert transforms along curves," *Bull. Amer. Math. Soc.*, **80** (1974), 106–108.
10. Sally, P., "Harmonic analysis and group representations," this volume.
11. Stein, E. M., *Singular Integrals and Differentiability Properties of Functions*, Princeton University Press, Princeton, 1970.

---

*For some more background see the paper [5].

12. ——, "Some problems in harmonic analysis suggested by symmetric spaces and semi-simple Lie groups," *Proc. Inter. Congress of Math.*, **1** (1970), 173–189.

13. ——, and G. Weiss, *Introduction to Fourier Analysis on Euclidean Spaces*, Princeton University Press, Princeton, 1971.

14. Tomas, P., "A restriction theorem for the Fourier transform," *Bull. Amer. Math. Soc.*, (to appear).

15. Wiener, N., *The Fourier Integral and Certain of its Applications*, Cambridge University Press, Cambridge, 1973, 46–57.

16. Zygmund, A., *Trigonometric Series*, (2nd edition), 2 vols., Cambridge University Press, Cambridge, 1959.

17. ——, "On Fourier coefficients and transforms of functions of two variables," *Studia Math.*, **50** (1974), 189–201.

HARMONIC ANALYSIS AND PROBABILITY

D. L. Burkholder

Probability theory is often used in harmonic analysis to construct examples and counterexamples. There is a long and rich history of this and the names of Paley, Zygmund, Wiener, and many others come to mind. Widely appreciated, this kind of application of probability is flourishing now more than ever before.

The emphasis here, however, will be on a different kind of application: the role of probability in the construction of proofs. A theorem may have no apparent connection with probability and yet the ideas and results of probability theory can sometimes be used directly or indirectly in the proof. Even if another proof is already known, the probabilistic proof may give additional insight.

Our first example of this is Burgess Davis's proof [5] of Kolmogorov's weak-type inequality for the conjugate function [12]. Although many good proofs exist, Davis's proof has the advantage of giving the best constant.

The Kolmogorov inequality is a consequence of the following inequality for analytic functions. Let **C** be the complex plane and

$F = u + iv$ a function analytic in $\mathbf{C}$ with $v(0) = 0$. (It would be enough to assume that F is a polynomial.) Define the real functions f and $\tilde{f}$ on $[0, 2\pi)$ by $f(\theta) = u(e^{i\theta})$ and $\tilde{f}(\theta) = v(e^{i\theta})$. Then, for all positive real numbers λ,

$$m(|\tilde{f}| \geqslant \lambda) \leqslant c\lambda^{-1} \int_0^{2\pi} |f(\theta)|\, d\theta, \tag{1}$$

where the left-hand side denotes the Lebesgue measure of $\{\theta \in [0, 2\pi) : |\tilde{f}(\theta)| \geqslant \lambda\}$ and the choice of the real number c does not depend on F or λ.

Burgess Davis's beautiful and intuitive proof of this uses Brownian motion.

Before defining Brownian motion, we consider a measure space much richer in its structure than the Lebesgue space $[0, 2\pi)$: Wiener space. The points of this new space Ω are the continuous functions $\omega : [0, \infty) \to \mathbf{C}$ with $\omega(0) = 0$. (Think of a traveler in the complex plane who starts at the origin and has position $\omega(t)$ at time t.) What are the measurable sets of Ω? If $t \geqslant 0$, let $Z_t : \Omega \to \mathbf{C}$ be defined by $Z_t(\omega) = \omega(t)$. At the very least, all sets of the form $\{Z_t \in B\}$ should be measurable, where B is a Borel set of $\mathbf{C}$. So the collection of measurable sets is taken to be simply the smallest σ-field $\mathcal{Q}$ containing all such sets. Relative to $\mathcal{Q}$, Z_t is a measurable function.

Wiener [17] proved the existence of a particularly interesting measure for $(\Omega, \mathcal{Q})$: *There exists a measure P on $\mathcal{Q}$ such that*

(i) *if $t > 0$ and B is a Borel set, then*

$$P(Z_t \in B) = \frac{1}{2\pi t} \int\int_B \exp\left(-\frac{|z|^2}{2t}\right) dz;$$

(ii) *if $s > 0$, $t > 0$, and B is a Borel set, then*

$$P(Z_{t+s} - Z_s \in B) = P(Z_t \in B);$$

(iii) *if n is a positive integer, $0 \leqslant s_1 < t_1 \leqslant \cdots \leqslant s_n < t_n$, and $B_1, \ldots, B_n$ are Borel sets, then*

$$P(Z_{t_1} - Z_{s_1} \in B_1, \ldots, Z_{t_n} - Z_{s_n} \in B_n) = \prod_{j=1}^{n} P(Z_{t_j} - Z_{s_j} \in B_j).$$

So Z_t has a Gaussian distribution; (ii) is the property of stationary increments (the traveler never gets tired); (iii) is the property of independent increments (the traveler has no memory).

We are now ready to define Brownian motion. Consider any family $Z = \{Z_t, t \geqslant 0\}$ of measurable complex functions on a probability space $(\Omega, \mathcal{A}, P)$. If Z starts at the origin ($Z_0 = 0$), has continuous sample functions $t \to Z_t(\omega)$ and satisfies (i), (ii), and (iii), then Z is a (complex) *Brownian motion* or *Wiener process*.

To prove Kolmogorov's inequality for the conjugate function, we need just a few rather simple properties of Brownian motion.

The first is *rotational invariance*: *If Z is a Brownian motion and $\theta \in [0, 2\pi)$, then $e^{i\theta}Z$ is a Brownian motion*. Properties (ii) and (iii) clearly hold for $e^{i\theta}Z$ and (i) holds by the rotational invariance of the integrand.

Consider one of the consequences of rotational invariance. Define $\mu : \Omega \to [0, \infty]$ by

$$\mu(\omega) = \inf\{t : |Z_t(\omega)| = 1\}$$

where $\inf \phi = \infty$. Then μ, the first time that Z hits the unit circle, is finite with probability one: Using (i), we have that

$$\begin{aligned} P(\mu = \infty) &\leqslant P(\mu > t) \leqslant P(|Z_t| < 1) \\ &= \frac{1}{2\pi t} \iint_{|z|<1} \exp\left(-\frac{|z|^2}{2t}\right) dz \\ &\leqslant \frac{1}{2\pi t} \iint_{|z|<1} dz = \frac{1}{2t}, \end{aligned}$$

which approaches zero as $t \to \infty$. Let Z_μ be defined on $\{\mu < \infty\}$ by $Z_\mu(\omega) = Z_{\mu(\omega)}(\omega)$. Then Z_μ, the hitting position of the Brownian motion Z on the unit circle, is defined essentially everywhere. Now let A be an arc of the unit circle. By rotational invariance,

$$\begin{aligned} P(Z_\mu \in A) &= P(e^{i\theta} Z_\mu \in A) \\ &= P(Z_\mu \in e^{-i\theta} A), \end{aligned}$$

so $P(Z_\mu \in A)$ depends only on the length of A and not its position. In particular,

$$P(Z_\mu \in A) = |A|/2\pi,$$

where $|A|$ denotes the arclength of A. This implies that if B is a Borel subset of $[0, 2\pi)$ and $\arg Z_\mu$ is here understood to have its values in $[0, 2\pi)$, then

$$P(\arg Z_\mu \in B) = m(B)/2\pi. \tag{2}$$

How can we exploit the fact that the Lebesgue measure space $[0, 2\pi)$ appears so naturally in this Brownian motion setting? To prove (1), we can take $\lambda = 1$. So consider $B = \{\theta \in [0, 2\pi) : |\tilde{f}(\theta)| \geqslant 1\}$. By (2),

$$\begin{aligned} P(|v(Z_\mu)| \geqslant 1) &= P(\arg Z_\mu \in B) \\ &= m(B)/2\pi \\ &= m(|\tilde{f}| \geqslant 1)/2\pi. \end{aligned}$$

Also, by (2), we have that

$$\begin{aligned} E|u(Z_\mu)| &= E|f(\arg Z_\mu)| \\ &= \int_0^{2\pi} |f(\theta)|\, d\theta/2\pi, \end{aligned}$$

where E, as usual, denotes integration over Ω with respect to P. So, by rotational invariance, the proof of Kolmogorov's inequality for the conjugate function reduces to proving

$$P(|v(Z_\mu)| \geqslant 1) \leqslant cE|u(Z_\mu)|. \tag{3}$$

The second property of Brownian motion that we need is a little more subtle but is still quite intuitive. Its proof is also based on the symmetry properties of Brownian motion. Write $Z_t = X_t + iY_t$ where X_t and Y_t are real. Let

$$\tau(\omega) = \inf\{t : |Y_t(\omega)| = 1\}.$$

Like μ defined above, τ is a stopping time of Z, that is, τ maps Ω into $[0, \infty]$ and, if $t > 0$, then $\{\tau < t\} \in \mathcal{Q}_t$ where $\mathcal{Q}_t$ is the smallest sub-σ-field of $\mathcal{Q}$ with respect to which every Z_s is measurable, $0 \leqslant s \leqslant t$. Now let ν be any stopping time of Z such that $\{X_{\nu \wedge t}, t \geqslant 0\}$ is uniformly integrable, that is, such that

$$\sup_{t \geqslant 0} \int_{\{|X_{\nu \wedge t}| > \lambda\}} |X_{\nu \wedge t}| \, dP \to 0$$

as $\lambda \to \infty$, where $\nu \wedge t = \min\{\nu, t\}$. (Uniform integrability is clearly implied by uniform boundedness, which holds in our first application; it is also implied by $E\nu < \infty$, for example, which holds in our second application.) Under this condition on ν, Burgess Davis [5] proved that

$$P(|Y_\nu| \geqslant 1) \leqslant E|X_\nu| / E|X_\tau|.$$

Note that both sides equal one if $\nu = \tau$. This inequality seems quite intuitive: If $E|X_\nu|$ is small compared to $E|X_\tau|$, it seems plausible that ν is usually smaller than τ and $P(|Y_\nu| \geqslant 1)$ is small compared to $P(|Y_\tau| \geqslant 1) = 1$. But some condition on ν is clearly necessary to rule out stopping times such as the first time Z hits the imaginary axis outside the unit disc; in this case, $P(|Y_\nu| \geqslant 1) = 1$ but $E|X_\nu| = 0$. Let $K = 1/E|X_\tau|$. It is not hard to show that

$$K = \frac{1 + 3^{-2} + 5^{-2} + 7^{-2} + \cdots}{1 - 3^{-2} + 5^{-2} - 7^{-2} + \cdots}. \tag{4}$$

We summarize:

LEMMA 1 (Davis [5]): *If $Z = X + iY$ is a complex Brownian motion and ν is a stopping time of Z such that $\{X_{\nu \wedge t}, t \geqslant 0\}$ is uniformly integrable, then*

$$P(|Y_\nu| \geqslant 1) \leqslant KE|X_\nu|$$

and the inequality is sharp.

Conformal invariance is the third property of Brownian motion that we need. This is a strengthening of rotational invariance and was first noted by Paul Lévy [13]: roughly, an analytic function maps Brownian motion into Brownian motion. (Maps preserving angles must also preserve the wiggles in the typical path of a Brownian traveler.)

We shall now be precise. Let F be analytic in $\mathbf{C}$ as before. It is slightly more convenient to assume that not only $v(0) = 0$ but also that $u(0) = 0$, and we shall do so from now on. Also, we can assume that F is nonconstant. Then $\{F(Z_t), t \geqslant 0\}$ starts at the origin, has continuous paths, and satisfies (iii), but not necessarily (i) and (ii). However, with a change of time, this process does become Brownian motion: There is an increasing continuous (random) function $\alpha : [0, \infty) \to [0, \infty)$ such that W defined by $W_t = F(Z_{\alpha(t)})$, $t \geqslant 0$, is a Brownian motion. Let

$$\beta(t) = \int_0^t |F'(Z_s)|^2 \, ds, \qquad t \geqslant 0; \tag{5}$$

then, with probability one, β is continuous and increasing and α is the inverse of β.

Let μ be the stopping time in (3) and let $\nu = \beta(\mu)$. Then ν is (effectively) a stopping time of $W = U + iV$ and $F(Z_\mu) = W_\nu$. Therefore,

$$u(Z_\mu) = U_\nu, \qquad v(Z_\mu) = V_\nu$$

and (3), the inequality to be proved, becomes

$$P(|V_\nu| \geqslant 1) \leqslant cE|U_\nu|. \tag{6}$$

But (6) is true with $c = K$ by Lemma 1 applied to the Brownian motion W. Therefore (1) holds with $c = K$.

Remarks. In the above discussion, we have assumed for the sake of simplicity that F is analytic in the entire complex plane. This is not necessary. Suppose merely that F is analytic in $|z| < 1$ and $F(0) = 0$. Then, provided the underlying probability space is slightly enriched in a standard way, there is a Brownian motion W such that

$$F(Z_t) = W_{\beta(t)} \tag{7}$$

for $t < \mu$ where β is defined by (5) for $t \leqslant \mu$. In particular, consider the function

$$F(z) = \frac{2}{\pi} \log \frac{1+z}{1-z}$$

analytic in $|z| < 1$ and continuous on the closed unit disc except at $z = \pm 1$. Then, with probability one, (7) holds also for $t = \mu$. In (6), $P(|V_\nu| \geqslant 1) = 1$ and $E|U_\nu| = 1/K$ since F maps $|z| < 1$ onto the strip

$$\{w \in \mathbf{C} : -1 < \operatorname{Im} w < 1\}$$

so the stopping time ν is here the same as

$$\tau = \inf\{t : |V_t| = 1\}.$$

So K is the smallest possible constant c in (6). This implies that K is the best constant in Kolmogorov's inequality.

In Lemma 1, the inequality still holds if $E|X_\nu|$ is replaced by $E|X_\nu + x|$, where x is any real number. This fact makes it possible to drop the assumption, made above for simplicity, that $u(0) = 0$.

The number K of (4) is also the best constant in the weak-type (1, 1) inequality for the Hilbert transform (see [5]).

A uniqueness theorem. Brownian motion makes it possible to gain a deeper understanding of many other classical results. Consider the following result due to Privalov: *Let F be analytic in $|z| < 1$ and have the nontangential limit zero in a set of positive measure on $|z| = 1$. Then F is identically zero in $|z| < 1$.* For the classical approach to this theorem, see Chapter 14 of [19]. For a probabilistic proof of a more general theorem, see Doob [7]. Our proof here of Privalov's theorem is similar to Doob's proof but is perhaps a little more direct and self-contained.

We need another basic property of Brownian motion: *If Z is a complex Brownian motion and z is a nonzero complex number, then*

$$P(Z_t = z \text{ for some } t \geqslant 0) = 0, \tag{8}$$

$$P(Z_t \to z \text{ as } t \to \infty) = 0. \tag{9}$$

The proof of (9) uses an argument that we have already used: By (i), the probability in (9) does not exceed

$$\lim_{t\to\infty} P(|Z_t| < |z| + 1) = 0.$$

There are several different proofs of (8); see [10] and [6]. Here we give a proof using conformal invariance discovered recently by Burgess Davis. By conformal invariance, the process

$$\{z(1 - \exp Z_t), t \geqslant 0\}$$

is a Brownian motion up to change in time. Therefore, the probability in (8) equals

$$P(z(1 - \exp Z_t) = z \quad \text{for some} \quad t \geqslant 0).$$

But since the exponential function never vanishes, this set is empty and has probability zero.

Let $\mu = \inf\{t : |Z_t| = 1\}$ as before. We show below that the hypothesis of the Privalov theorem implies that

$$P(F(Z_t) \to 0 \quad as \quad t \to \mu) > 0 \tag{10}$$

where the restriction $t < \mu$ is understood.

Now suppose that F is not identically zero. Then we may assume without loss of generality that $F(0) \neq 0$. Applying (7) to $F(0) - F(z)$, we have the existence of a Brownian motion W such that

$$F(0) - F(Z_t) = W_{\beta(t)}$$

for $t < \mu$ where β is defined by (5) for $t \leqslant \mu$. Let $\nu = \beta(\mu)$. Then

(10) implies that

$$P(W_t \to F(0) \quad \text{as} \quad t \to \nu) > 0.$$

But this is impossible, as we can see by considering the two cases $\nu < \infty$ and $\nu = \infty$, and applying (8) and (9) to W and $z = F(0)$. Therefore, F is identically zero.

We now prove (10). For $0 < a < 1$, let $\Gamma_a(\theta)$ be the interior of the smallest convex set containing the point $e^{i\theta}$ and the circle $|z| = a$. If $h > 0$, let

$$\Gamma_{a,h}(\theta) = \Gamma_a(\theta) \cap \{z : 1 - h < |z| < 1\}$$

and $N_{a,h}(\theta) = \sup\{|F(z)| : z \in \Gamma_{a,h}(\theta)\}$. Fix a and let

$$B_a = \left\{\theta \in [0, 2\pi) : \lim_{h \to 0} N_{a,h}(\theta) = 0\right\}.$$

By the hypothesis of Privalov's theorem, $m(B_a) > 0$. It was shown in [4] that

$$P\left(\sup_{0 \leqslant t < \mu} |F(Z_t)| > \lambda\right) \leqslant c_a m(N_a > \lambda), \quad \lambda > 0,$$

where $N_a = N_{a,1}$. Here we shall slightly modify the proof of this inequality to obtain the following local result: *If B is a Borel set of $[0, 2\pi)$, then*

$$P\left(\limsup_{t \to \mu} |F(Z_t)| > \lambda, \quad \arg Z_\mu \in B\right)$$

$$\leqslant c_a m\{\theta \in B : N_{a,h}(\theta) > \lambda\} \quad (11)$$

for all $\lambda > 0$ and $h > 0$, where the choice of the real number c_a depends only on a. Substituting B_a for B and letting $h \to 0$ gives

$$P\left(\limsup_{t \to \mu} |F(Z_t)| > \lambda, \quad \arg Z_\mu \in B_a\right) = 0.$$

Now let $\lambda \to 0$ to see that the probability in (10) must be at least as large as

$$P(\arg Z_\mu \in B_a) = m(B_a)/2\pi > 0.$$

Thus, we need only verify (11) to complete the proof of Privalov's theorem.

Each side of (11) is a regular measure of the Borel sets B of $[0, 2\pi)$. Therefore, we may replace B by a set $G \subset [0, 2\pi)$ such that $e^{iG} = \{e^{i\theta} : \theta \in G\}$ is open in $|z| = 1$. Let $H = \{N_{a,h} > \lambda\} \cap G$. Then e^{iH} is also open in $|z| = 1$. Let V be the interior of the complement of $\cup\{\Gamma_{a,h}(\theta) : \theta \not\in H\}$ relative to the set $1 - h < |z| < 1$ where we may suppose that $0 < h < 1 - a$. Then, V has the familiar saw-toothed shape (see page 200 of Volume II of [19] or Fefferman's paper in this volume) slightly modified: each tooth of height greater than h has been filed down to height h. Note that the closure of e^{iH} is that part of the boundary of V in $|z| = 1$. Denote the other part by

$$\partial_0 V = \{z \in \partial V : |z| < 1\}.$$

Let $\tau = \inf\{t : Z_t \in \partial_0 V \text{ or } |Z_t| = 1\}$. Then $\{Z_\tau \in \partial_0 V\} = \{\tau < \mu\}$ and $Z_\mu \in e^{iH}$ implies that $\tau < \mu$. Therefore,

$$\begin{aligned} m(N_{a,h} > \lambda, G) &= m(H) \\ &= 2\pi P(Z_\mu \in e^{iH}) \\ &= 2\pi P(\tau < \mu, Z_\mu \in e^{iH}). \end{aligned}$$

If $z = re^{i\theta}$ is in the open unit disc, then

$$p(z, H) = \frac{1}{2\pi} \int_H \frac{1 - r^2}{1 - 2r\cos(\theta - t) + r^2}\, dt$$

is the probability that a Brownian motion starting from z hits e^{iH} before the other part of $|z| = 1$. This is due to Kakutani [10]. Moreover, if $z \in \partial_0 V$, then $p(z, H) \geqslant c_a$, where the choice of the positive real number c_a depends only on a; see Theorem 6.18 on page 99 of Volume I of [19]. By the strong Markov property of Brownian motion (see [14], for example), we have that

$$P\left(\tau < \mu, Z_\mu \in e^{iH}\right) = \int_{\{\tau<\mu\}} p(Z_\tau, H)\, dP$$
$$\geqslant c_a P(\tau < \mu).$$

Finally, since e^{iG} is open in $|z| = 1$,

$$\left\{ \limsup_{t\to\mu} |F(Z_t)| > \lambda, \quad Z_\mu \in e^{iG} \right\} \subset \{\tau < \mu\}.$$

To see this, let ω belong to the left-hand side. Then there is a positive number $t < \mu(\omega)$ such that $Z_t(\omega)$ belongs to the saw-toothed region determined by G and $|F(Z_t(\omega))| > \lambda$. Therefore, $Z_t(\omega) \in V$ so, by the definition of τ, $\tau(\omega) < t < \mu(\omega)$ and ω belongs to the right-hand side. Combining these estimates, we obtain (11) with a new choice for c_a. This completes the proof of Privalov's theorem.

Summary and further remarks. To illustrate some of the possible applications of probability theory, we have given Brownian motion proofs of two classical theorems: Kolmogorov's weak-type inequality for the conjugate function and Privalov's uniqueness theorem.

Many other similar examples could have been included.

Indeed, a number of theorems in harmonic analysis were first proved with the use of probability. In fact, in the original proof of his inequality for the conjugate function, Kolmogorov used stochastic independence in an essential way. Several "classical" proofs were discovered later.

This experience is typical. Only rarely does a probabilistic proof stand alone for long.

In this connection, we suggest the following problem: Prove without using probability that the number K in (4) is a permissible choice for the constant in the Kolmogorov inequality. (Added in proof: Albert Baernstein has recently been able to do this.) It is easy to show without using any probability that no smaller number than K is permissible: calculate both sides of (1) for f and $\tilde{f}$ corresponding to

$$F(z) = \frac{2}{\pi} \log \frac{1+z}{1-z} .$$

Pichorides, using classical methods, has found the best constants for the strong L^p inequalities of M. Riesz for the conjugate function [16]. A Brownian motion proof of these inequalities is known (see page 149 of [4]) but it would be interesting to have such a proof giving the best constants.

It should be made clear that probabilistic proofs are often no simpler than classical proofs from a technical point of view. The advantage of a good probabilistic proof is not that the technical details become easier, although this is sometimes the case, but that the underlying ideas become more transparent. For example, the truth of Privalov's theorem becomes evident once we see that it is a question of whether or not Brownian motion hits a particular point in the complex plane with positive probability.

Sometimes probability plays a role even if a proof makes no apparent use of it. For example, in [3], no probability is used or mentioned. Yet both the theorems and proofs contained therein were strongly motivated by the analogous probability ideas explored earlier in [2].

Further reading. There are many excellent sources for interesting applications of probability theory to the construction of examples and counterexamples. Here is a small sample: Zygmund [19, pages 212–222 of Volume I], Kahane [8, 9], and Kaufman [11].

For a good short introduction to Brownian motion, see McKean [14]. In particular, the existence of Brownian motion is proved (pages 5–8) using the elegant method of Lévy and Ciesielski and its conformal invariance is established rigorously (page 109) using the Itô integral. For an excellent discussion of the historical and physical background, see Nelson [15].

Here we have illustrated the contributions that probability theory can make to harmonic analysis and have entirely neglected the applications of harmonic analysis to probability. These are better known and some can be found in any advanced text on probability theory; in this connection, the monograph of Bochner [1] should not be overlooked. Several of the major works of Wiener contain a nice combination of the two points of view; see [18], for example.

REFERENCES

1. Bochner, S., *Harmonic Analysis and the Theory of Probability*, Berkeley: University of California Press, 1960.
2. Burkholder, D. L., and R. F. Gundy, "Extrapolation and interpolation of quasi-linear operators on martingales", *Acta Math.*, **124** (1970), 249–304.
3. ——, "Distribution function inequalities for the area integral", *Studia Math.*, **44** (1972), 527–544.
4. Burkholder, D. L., Gundy, R. F., and M. L. Silverstein, "A maximal function characterization of the class H^p," *Trans. Amer. Math. Soc.*, **157** (1971), 137–153.
5. Davis, B., "On the weak type (1, 1) inequality for conjugate functions," *Proc. Amer. Math. Soc.*, **44** (1974), 307–311.
6. Doob, J. L., "Semimartingales and subharmonic functions," *Trans. Amer. Math. Soc.*, **77** (1954), 86–121.
7. ——, "Conformally invariant cluster value theory," *Illinois J. Math.*, **5** (1961), 521–549.
8. Kahane, J.-P., *Some Random Series of Functions*, Lexington, Massachusetts: Heath, 1968.
9. ——, "The technique of using random measures and random sets in harmonic analysis," *Advances in Probability and Related Topics*, edited by Peter Ney, **1** (1971), 65–101, New York: Marcel Dekker, Inc.
10. Kakutani, S., "Two-dimensional Brownian motion and harmonic functions," *Proc. Imperial Acad. Japan*, **20** (1944), 706–714.
11. Kaufman, R., "Fourier analysis and paths of Brownian motion," *Bull. Soc. Math. France*, to appear.

12. Kolmogorov, A. N., "Sur les fonctions harmoniques conjuguées et les séries de Fourier," *Fund. Math.*, **7** (1925), 24–29.

13. Lévy, P., *Processus stochastiques et mouvement brownien*, Paris: Gauthier-Villars, 1948, new edition 1965.

14. McKean, H. P., Jr., *Stochastic Integrals*, New York: Academic Press, 1969.

15. Nelson, E., *Dynamical Theories of Brownian Motion*, Princeton: Princeton University Press, 1967.

16. Pichorides, S. K., "On the best values of the constants in the theorems of M. Riesz, Zygmund and Kolmogorov," *Studia Math.*, **44** (1972), 165–179.

17. Wiener, N., "Differential-space," *J. Math. and Phys.*, **2** (1923), 131–174.

18. Wiener, N., "Generalized harmonic analysis," *Acta Math.*, **55** (1930), 117–258.

19. Zygmund, A., *Trigonometric Series*, Cambridge: Cambridge University Press, 1959.

HARMONIC ANALYSIS OF MEAN-PERIODIC FUNCTIONS

Yves Meyer

For complex-valued functions on the line, a large class of functional equations may be defined generalizing the ordinary differential equations with constant coefficients. Solutions of these equations may be expanded into generalized Fourier series leading to a new kind of harmonic analysis. Generalization to n dimensional spaces will be given and open problems will be discussed.

1. A CLASSICAL EXAMPLE

Assume you are given an ordinary differential equation, for example

$$y''' - 3y' + 2y = 0. \tag{1}$$

To solve it, you first look at elementary solutions $y(x) = e^{rx}$, where r may be a complex number; you are led to the characteristic equation $r^3 - 3r + 2 = (r - 1)^2(r + 2) = 0$. If a root r presents some multiplicity m (here $r = 1$ implies $m = 2$), each $y(x)$

$= P(x)e^{rx}$ where P is a polynomial whose degree does not exceed $m - 1$ will be a solution. Then, adding what you just found you get all solutions; in our case, they are

$$y(x) = (ax + b)e^{x} + ce^{-2x}.$$

If you are given some initial data like $y(0)$, $y'(0)$ and $y''(0)$, then a, b, and c can be determined and y is fully specified.

2. THE PERIODIC CASE

Take now

$$f(x + 1) = f(x) \tag{2}$$

where f is a complex-valued continuous function on the line. It is a surprising fact that the way of solving (2) is the same as that for (1). Elementary solutions of the type e^{rx} are now $e^{2\pi inx}$, $n = 0, \pm 1, \ldots$. Adding up these exponential solutions, finite sums

$$S(x) = \Sigma\, a_n e^{2\pi inx} \tag{3}$$

are obtained. But now you have to do something else since, for example, $|\sin \pi x|$ is a continuous solution of (2) and cannot be a finite sum (3) since it is not differentiable at 0. To get all continuous solutions of (2) it is in fact enough to take all limits of all sequences $S_k(x)$ of type (3) which are uniformly convergent on $[0, 1]$.

Of course there are systematic ways of approaching a given solution of (2) by employing those sums of type (3) called the Cesáro means of the Fourier series of f, but we shall not talk about this since these ways do not generalize to other equations.

3. DIFFERENTIAL EQUATIONS WITH DIFFERENCES

We combine now example (1) with (2). Consider, for example,

$$f(x + \sqrt{2} + 1) = \tfrac{1}{3} f(x + \sqrt{2}\,) - \tfrac{1}{3} f(x + 1) + f(x), \tag{4}$$

or

$$f'(x + \sqrt{2} + 1) = f(x + \sqrt{2}\,) - f(x + 1) + f'(x), \tag{5}$$

or any other equation of the type

$$\sum_{1 \leqslant j \leqslant n} a_j \left(\frac{d}{dx} \right)^{k_j} f(x - x_j) = 0. \tag{6}$$

To solve (4), (5) or (6), you follow exactly the same paths as for (1) and (2). We don't state that as a theorem since a more general statement will be given in a moment. You seek all exponential solutions e^{rx} times polynomials whose degrees do not exceed the multiplicity minus one of the root r. You add up these elementary solutions. You take all limits of what you just obtained. But let us be more specific about this notion of multiplicity of a root and about the notion of limit you need.

In seeking solutions of (4) of type e^{rx} you are led to

$$F(r) = 3e^{(\sqrt{2}+1)r} - e^{\sqrt{2}\,r} + e^{r} - 3 = 0. \tag{7}$$

Now $F(r)$ is an entire function in the complex plane with an infinite number of zeros. Write all zeros of F as a sequence r_1, $r_2, \ldots, r_k, \ldots,$ (usually they are ordered with increasing moduli). The multiplicity of a root of an entire function makes sense. So we write r_k m_k-times in our sequence if the multiplicity of the root r_k is m_k ($r_{k+1} = r_k, \ldots, r_{k+m_k-1} = r_k$); in that case

$$e^{r_k x}, \qquad x e^{r_k x}, \ldots, x^{m_k - 1} e^{r_k x}, \tag{8}$$

are still solutions of (4). We form all finite sums of these elementary solutions

$$S_j(x) = \Sigma P_k(x) e^{r_k x} \tag{9}$$

where the degree of $P_k \leqslant m_k - 1$, and to finish the job we take all possible limits

$$f(x) = \lim_{j \to +\infty} S_j(x), \tag{10}$$

where the limit is uniform on each compact interval $[-T, T]$. We have obtained all solutions of (4). For the equation (5), we have to replace (10) by

$$f(x) = \lim_{j \to +\infty} S_j(x) \quad \text{and} \quad f'(x) = \lim_{j \to +\infty} S_j'(x) \tag{11}$$

and with the same steps we get all continuous solutions of (5) with a continuous derivative.

To end with these two examples we have to show what are "initial data" in these cases. In the case of (4) it may be easily shown that if you are given on $[0, 1 + \sqrt{2}\,]$ any continuous function $g(x)$ such that

$$g(\sqrt{2} + 1) = \tfrac{1}{3}\, g(\sqrt{2}\,) - \tfrac{1}{3}\, g(1) + g(0)$$

then g extends to a unique solution f of (4); this statement is similar to the fact that any continuous function on $[0, 1]$ satisfying $f(0) = f(1)$ may be extended to a periodic function of period 1. This extension f may be written explicitly in terms of g and we obtain the estimate

$$|f(x)| \leqslant C|x|^{1/2} \sup_{[0,\, 1+\sqrt{2}\,]}|f| \tag{12}$$

where C is some absolute constant. Similar statements are valid for (5) and (6).

4. CONVOLUTION EQUATIONS

In order to generalize equations (4) and (5) we need to be a little more systematic. Let $k \geqslant 0$ be an integer and $\mathcal{C}^k(\mathbf{R})$ the vector space of all complex-valued functions f on the line such that f, $f', \ldots, f^{(k)}$ are continuous. A sequence f_j, $j \geqslant 0$, in $\mathcal{C}^k(\mathbf{R})$ is said to converge to zero if f_j and all its derivatives up to the order k are uniformly convergent to 0 on each closed interval $[-T, T]$.

A finitely supported measure μ is a mathematical object assigning given masses a_j, $1 \leqslant j \leqslant n$, to given points x_j on the line (these masses may be complex numbers). A "test function" will be any function f in $\mathcal{C}(\mathbf{R})$ and the scalar product between f and μ is defined to be $\Sigma_1^n\, a_j f(x_j)$ and is written $\int f\, d\mu$. We define a convolution product between f and μ by the rule

$$(f * \mu)(x) = \int f(x - t)\, d\mu(t).$$

In our case we obtain $\Sigma_1^n a_j f(x - x_j)$. Now any equation of type (4) may be written in the form

$$f * \mu = 0. \tag{13}$$

A Radon measure with compact support on the line is a continuous linear form over $\mathcal{C}(\mathbf{R})$. If μ is such a measure, (13) makes sense and is the correct generalization of (4). To generalize (5) we need the concept of a distribution S with compact support. For example the rule

$$\langle S, f \rangle = f(0) - 2f'(1) \tag{14}$$

defines a continuous linear form over test functions f in $\mathcal{C}^1(\mathbf{R})$. A distribution S of type k, $k \geqslant 0$, is a continuous linear form over $\mathcal{C}^k(\mathbf{R})$ (the continuity means that every time a sequence f_j in $\mathcal{C}^k(\mathbf{R})$ tends to 0, $S(f_j)$ should tend to 0). We say that a distribution S is supported in the compact interval $[-l, l]$ if $S(f) = 0$ whenever the test function vanishes on $[-l, l]$ (or on some neighborhood of 0 if $l = 0$). Convolution between S of type k and a test function f in $\mathcal{C}^k(\mathbf{R})$ is a continuous function on the line whose value at x is the value of the linear form S at the function $t \to f(x - t)$. In case S is given by (14), $S * f = f(x) + 2f'(x - 1)$. Now we may see that an equation of the type

$$S * f = 0 \tag{15}$$

where S is a given distribution of type k and f is an unknown function in $\mathcal{C}^k(\mathbf{R})$ is the correct generalization of (5). Of course, S is assumed to be distinct from the 0 distribution and then any solution of (15) is, by definition, a mean-periodic function.

5. COMPLEX FOURIER TRANSFORMS AND PALEY-WIENER'S THEOREM

Let S be a distribution of type k supported in $[-l, l]$. Then, for each fixed complex number z, $t \to e^{-itz} = e_z(t)$ is a test function in $\mathcal{C}^k(\mathbf{R})$. Our distribution S is a linear form over such test functions and the value of S at e_z is, by definition, the complex

Fourier transform of S at z and is written $\hat{S}(z)$. Then $\hat{S}(z)$ is an entire function on the complex plane. Moreover, for each $\varepsilon > 0$, a constant $A(\varepsilon)$ may be found such that, for each z in the complex plane, we have

$$|\hat{S}(z)| \leqslant A(\varepsilon)e^{(l+\varepsilon)|z|} \tag{16}$$

and there is another constant B and an integer n such that, for each real x

$$|\hat{S}(x)| \leqslant B(1 + x^2)^n. \tag{17}$$

But the converse is also true as it was proved by Paley and Wiener. Any entire function F on the complex plane fulfilling (16) and (17) is the Fourier transform of a distribution supported on $[-l, l]$.

6. SCHWARTZ'S THEOREM

THEOREM 1: *Let S be a given distribution of type k supported on the compact interval $[-l, l]$. Consider the convolution equation*

$$f * S = 0 \tag{18}$$

in which the unknown function f belongs to $\mathcal{C}^k(\mathbf{R})$.

To find all solutions f in $\mathcal{C}^{(k)}(\mathbf{R})$ of (18) it is sufficient

(a) *to seek all solutions of type $f(x) = x^n e^{i\lambda x}$ whenever λ satisfies $\hat{S}(\lambda) = 0$ with multiplicity at least $n + 1$;*

(b) *to add all such elementary solutions;*

(c) *to take all limits, in $\mathcal{C}^k(\mathbf{R})$, of finite sums found in (b).*

Moreover, if two solutions coincide on a neighborhood of $[-l, l]$ supporting S, they coincide everywhere: "initial data" for (18) are values in some neighborhood of $[-l, l]$. Unfortunately it is generally impossible to express f outside $[-l, l]$ in terms of such initial data as was the case for the special equations (4), (5) and (6).

7. BEURLING-MALLIAVIN'S THEOREM

For applying theorem 1 it is interesting to have some idea about the distribution of those λ in the complex plane for which $e^{i\lambda x}$ is a solution of (18). When our distribution S is supported by 0, S is a finite sum of derivatives of Dirac masses concentrated at 0 and (18) is an ordinary linear differential equation with constant coefficients. In that case there are only a finite number of λ's. In the general case these λ are nicely distributed around the real axis and their "density" is $1/2\pi$ times the length of the smallest interval supporting S.

But this notion of a density has to be defined precisely. Let E be a set whose elements are open and finite intervals ω on the line. For each ω, call R_ω the interior of the square of the complex plane with diagonal ω. Let $R(E)$ be the (open) union of all these open squares. Then E is said to be negligible if

$$\iint_{R(E)} \frac{dx\,dy}{1 + x^2 + y^2} < +\infty. \tag{19}$$

If there is some bound for the lengths of ω, we have such a situation and (19) expresses the fact that large intervals in E are rare and far.

Now let $(u_k)_0^\infty$ be a sequence of real numbers without finite accumulation point. For each open and finite interval ω, let $N(\omega)$ be the number of k for which $u_k \in \omega$. Call $|\omega|$ the length of ω, take the average number $N(\omega)/|\omega|$. We say that our sequence has a finite density d in the sense of Beurling-Malliavin if the following is true: for each $\varepsilon > 0$, the set $E(\varepsilon)$ of open ω such that

$$\left| \frac{N(\omega)}{|\omega|} - d \right| \geqslant \varepsilon$$

is a negligible set of intervals. (This means that for very large ω it is rare that the average number of points of $(u_k)_0^\infty$ in ω would be far from d.)

THEOREM 2: *Let S be a distribution on the line with compact support and assume that $[-l, l]$ is the smallest closed interval supporting S. Let $(\lambda_k)_0^\infty$ be the sequence of λ's such that $e^{i\lambda x}$ is a solution of* (18). *Then if $\lambda_k = \sigma_k + i\tau_k$,*

$$\sum_0^\infty \frac{|\tau_k|}{|\lambda_k|^2} < +\infty. \tag{20}$$

If, whenever $\sigma_k \neq 0$,

$$u_k = \frac{|\lambda_k|^2}{\sigma_k}, \tag{21}$$

then $(u_k)_0^\infty$ has density l/π in the sense of Beurling and Malliavin.

By convention a λ in our sequence is written m times if the corresponding multiplicity of $\hat{S}$ is m and, except for this case, for distinct k we write distinct λ_k. In (21) enter only those λ_k for which $\sigma_k \neq 0$; those for which $\sigma_k = 0$ have only to satisfy (20). Conversely, any sequence $(z_k)_0^\infty$ of complex numbers satisfying (20) and the density condition is contained in the set of λ arising from solutions of an equation (18); S in (18) may be chosen to be supported on $[-l-\varepsilon, l+\varepsilon]$ if the Beurling and Malliavin density given by (21) is l/π.

8. THE n-DIMENSIONAL CASE

If S is a given distribution supported by a compact subset K of $\mathbf{R}^n$, equation (18) makes sense; the unknown f is now an indefinitely differentiable function on $\mathbf{R}^n$. Specializing to the case where S is supported by a single point, equation (18) is a linear partial differential equation with constant coefficients. In all these cases the analogue of Schwartz's theorem is true and was proved by Malgrange ([6]); elementary solutions are now products

$$f(x_1, \ldots, x_n) = P(x_1, \ldots, x_n) e^{i(z_1 x_1 + \cdots + z_n x_n)}. \tag{22}$$

So we get approximations of solutions of (18) which are defined on the whole of $\mathbf{R}^n$. But Malgrange's theorem has also a local version: if Ω is a convex open subset of $\mathbf{R}^n$ and if we are given a partial differential equation with constant coefficients

$$\sum_{(q_1, \ldots, q_n)} C(q_1, \ldots, q_n) \frac{\partial^{q_1 + \cdots + q_n}}{(\partial x_1)^{q_1} \cdots (\partial x_n)^{q_n}} f = 0 \qquad (23)$$

then for any solution f of (23) which is defined on Ω and indefinitely differentiable there is a limit of sums of solutions of (23) of the form (22). To see that some condition on Ω is needed, take $n = 2$, $\mathbf{R}^2 = \mathbf{C}$, Ω defined by $1 < |z| < 2$ and the equation

$$\frac{\partial f}{\partial x} + i \frac{\partial f}{\partial y} = 0. \qquad (24)$$

Now (24) characterizes holomorphic functions $f : \Omega \to \mathbf{C}$. Elementary solutions of (24) are entire function and limits of sums of elementary solutions extend to $|z| < 2$; a solution of (24), such as $1/z$, cannot be such a limit.

9. OPEN PROBLEMS

The modern approach to convolution equations (18) is to assume that we have some information about the growth of f at infinity. (For example we know that our solution f is bounded.) In return we wish to get a sharper approximation by finite sums f_k of elementary solutions (22): if our solution f is bounded it would be quite plausible that these f_k be uniformly bounded over $\mathbf{R}^n$. A simple example of such a problem is given when $n = 3$ with bounded solutions of

$$f + (1 + i) \frac{\partial^2 f}{\partial x^2} + (2 + i) \frac{\partial^2 f}{\partial y^2} + (3 - i) \frac{\partial^2 f}{\partial z^2} = 0. \qquad (25)$$

Bounded elementary solutions are

$$f(x, y, z) = e^{i(ax + by + cz)}, \tag{26}$$

where a, b and c are real numbers that satisfy

$$a^2 + b^2 - c^2 = 0$$

$$1 = a^2 + 2b^2 + 3c^2.$$

It is an open problem to know if any bounded solution f of (25) is a limit of a uniformly bounded sequence S_k of sums of elementary solutions (26); that means $S_k(x) \to f(x)$ uniformly on compact subsets of $\mathbf{R}^3$ and the existence of some constant C for which $|S_k(x)| \leqslant C$ for all k and x in $\mathbf{R}^3$.†

Beautiful problems related to these ones concern weighted approximations of entire functions of several complex variables by polynomials. Quite effective tools seem to be the $\bar{\partial}$ estimates in ([8]).

REFERENCES

1. Beurling, A., and P. Malliavin, "On the closure of characters and zeros of entire functions," *Acta Math.*, **118** (1967), 79–93.

2. Delsarte, J., "Les fonctions moyenne-périodiques," *J. Math. Pures Appl.*, Ser. 9, **14** (1935), 403–453.

3. Domar, Y., "Sur la synthèse harmonique des courbes de $\mathbf{R}^2$," *C. R. Acad. Sci. Paris*, **270** (1970), 875–878.

4. Ehrenpreis, L., "Mean periodic functions," *Amer. J. Math.*, **78** (1955), 292–328.

5. Kahane, J.-P., *Lectures on mean-periodic functions*, Tata Institute of Fundamental Research, Bombay (1959).

6. Malgrange, B., "Existence et approximation de solutions des équations aux dérivées partielles et des équations de convolution," *Ann. Inst. Fourier (Grenoble)*, **6** (1955), 271–355.

7. Schwartz, L., "Théorie générale des fonctions moyenne-périodiques," *Ann. of Math.*, **48** (1947), 857–929.

8. Sibony, N., "These" (to appear), Bât. 425, *Faculté des Sciences*, 91405 Orsay, France.

†Added in proof: This problem has just been solved by Y. Domar, Uppsala.

HARMONIC ANALYSIS AND LCA GROUPS

Colin C. Graham

INTRODUCTION

This article begins where the paper, "Harmonic Analysis," by Guido Weiss [17] in the third volume of MAA studies, leaves off. We have three parts.

The first part is an outline of fundamental results about locally compact abelian groups (LCAGs). Most results are stated without proof; in a few places, the proof of an assertion is suggested. The second part discusses some harmonic analysis problems on the real line, and covers a much narrower area than the first, but this area is one which has been very fruitful and important in the development of the entire subject; a number of the real-line theorems stated here require the use of general LCAGs in their proofs. The third part contains some suggestions for further reading.

PART ONE. GENERAL THEORY. SECTIONS 1–7.

In this part, we discuss examples of LCAGs (Section 1). Haar measure is discussed in Section 2, and other measures on G are discussed in 3. Multiplicative linear functionals on $L^1(G)$ and the Pontryagin duality theorem are in 4, and the principal structure theorem is given in 5. In 6 we characterize all idempotent measures, and in 7 we discuss measures which are invertible in the algebra of measures; this section contains a detailed construction of a non-trivial measure.

1. LOCALLY COMPACT ABELIAN GROUPS

A locally compact abelian group (LCAG) is an abelian group (usually written additively with identity 0 and often called "G") endowed with a Hausdorff topology such that the group operations $+: G \times G \to G((x, y) \to x + y)$ and $-1: G \to G(x \to -x)$ are continuous maps from $G \times G \to G$ and $G \to G$ respectively and such that each element $x \in G$ has a neighborhood basis consisting of compact sets.

An abelian group G with the discrete topology (every set is open) is locally compact since if $x \in G$, then $\{x\}$ is a compact neighborhood of x; for example, the integers form a discrete LCAG. Some other examples of LCAG's are: the real line $\mathbf{R}$ (with the usual topology), Euclidean n-space $\mathbf{R}^n$ (usual topology), the n-fold product group $\mathbf{Z}^n$ (discrete topology on all n-tuples of integers; the group operation is the set of coordinate-wise addition).

All of the preceding groups are non-compact. The circle group $\mathbf{T}$ is the most important compact group. One way to obtain $\mathbf{T}$ is this:

The map from $\mathbf{R}$ to the complex numbers $\mathbf{C}$ given by $x \to e^{ix}$ is continuous, and converts addition in $\mathbf{R}$ to multiplication in $\mathbf{C}$. If $x = 2\pi n$, for some integer, n, then $e^{ix} = e^{2\pi in} = 1$, the identity of the multiplicative group, $\mathbf{T}$, of complex numbers of modulus one. Since $2\pi\mathbf{Z} = \{2\pi n : n \in \mathbf{Z}\} \subseteq \mathbf{R}$, we see that $\mathbf{T} = \mathbf{R}/2\pi\mathbf{Z}$, and multiplication of elements in $\mathbf{T}$ is the "same" as addition modulo 2π in $\mathbf{R}$. When we take the n-fold product T^n of T with itself, we obtain the last of the so-called classical groups, which are $\mathbf{T}$, $\mathbf{T}^n$, $\mathbf{R}$, $\mathbf{R}^n$, $\mathbf{Z}$, and $\mathbf{Z}^n$, $n \geqslant 1$. $\mathbf{T}$ is, as we said above, called the "circle group." Its identity is the number one.

The method used to construct $\mathbf{T}$ can be generalized, as follows. When G is a LCAG and H is a subgroup of G, then G/H is a group and is a topological group if it is given the "quotient topology:" a set $U \subseteq G/H$ is open if and only if $\cup\{x + H: x + H \in U\}$ is open in G. If H is closed in G, then G/H is locally compact. (This was how we obtained $\mathbf{T}$.) When H is an open subgroup, then every coset of H is open, so the complement of H is open, and so H is also closed, and G/H is discrete.

Sometimes the topology in a LCAG can be changed, so that with the new topology we obtain a new LCAG. We may always give a LCAG G the discrete topology, for example. The new discrete LCAG is usually called G_d. A less trivial example starts with the "torus" group $\mathbf{T}^2$. Fix an irrational number ξ and let $H = \{(e^{i\xi x}, e^{ix}) : x \in \mathbf{R}\} \subseteq \mathbf{T}^2$. Then H is a line wrapped around T^2. Give $G = \mathbf{T}^2$ that topology which makes H open, and gives H its usual line topology. (The irrational "slope" is necessary for H to be an infinite line and not a closed loop.)

We now give another family of examples of LCAG's. Let G and H be any two LCAGs, and let $\mathrm{Hom}(G, H)$ be the set of *continuous* group homomorphisms from G to H. We give $\mathrm{Hom}(G, H)$ the "*compact-open*" topology: a neighborhood basis at $\gamma \in \mathrm{Hom}(G, H)$ consists of sets U of the form $U = U(C, V, \gamma) = \{\rho \in \mathrm{Hom}(G, H): \rho(x) - \gamma(x) \in V \text{ all } x \in C\}$ where $C \subseteq G$ is compact, $\gamma \in \mathrm{Hom}(G, H)$ and V is open in H. The identity, 0, of $\mathrm{Hom}(G, H)$ is the constant function which sends all of G onto the identity of H. The formula $(\rho + \gamma)(x) = \rho(x) + \gamma(x)$, $(x \in \mathrm{H})$, defines the sum $\rho + \gamma$ of two elements of $\mathrm{Hom}(G, H)$. The case $H = T$ is particularly important and it is not hard to see that $\mathrm{Hom}(G, T)$ is discrete if G is compact, and that $\mathrm{Hom}(G, T)$ is compact if G is discrete (in which case, the compact open topology is the topology of pointwise convergence), and that more generally, Hom(G, H) is a topological group (though it may not be locally compact). Finally, Hom (G, T) *is* a LCAG whenever G is a LCAG.

The following equalities (isomorphisms) of LCAG groups can be verified with varying degrees of difficulty (the value of $\gamma \in$ Hom(G, T) at $x \in G$ is given in parentheses):

$$\begin{aligned} \mathrm{Hom}(\mathbf{T}, \mathbf{T}) &= \mathbf{Z}, \left(n(x) = e^{inx}\right) \\ \mathrm{Hom}(\mathbf{R}, \mathbf{T}) &= \mathbf{R}, \left(y(x) = e^{ixy}\right) \\ \mathrm{Hom}(\mathbf{R}, \mathbf{R}) &= \mathbf{R}, \left(y(x) = yx\right) \\ \mathrm{Hom}(\mathbf{Z}, \mathbf{T}) &= \mathbf{T}, \left(x(n) = e^{inx}\right) \end{aligned} \tag{1.1}$$

and finally

$$\mathrm{Hom}(G \times H, \mathbf{T}) = \mathrm{Hom}(G, \mathbf{T}) \times \mathrm{Hom}(H, \mathbf{T}).$$

We can obtain a new group by giving $\mathbf{T}$ the discrete topology (this new group is called $\mathbf{T}_d$) and writing

$$\mathrm{Hom}(\mathbf{T}_d, \mathbf{T}).$$

This last group is compact. Yet another new compact group can be obtained by letting $\mathbf{Z}_p$ be the finite cyclic group of order $p \geqslant 2$, and letting G be the direct sum of countable many copies of $\mathbf{Z}_p$. (That is, G is the set of all sequences $(n_1, n_2, \ldots)$ of integers $0 \leqslant n_j \leqslant p - 1$, of which only a finite number are not zero.) Addition in G is coordinate-wise modulo p, and G has the discrete topology. Our new group, called $\mathbf{D}_p$, is given by $\mathbf{D}_p = \mathrm{Hom}(G, \mathbf{T})$.

Since every element of $\mathbf{D}_p$ has order p, $\mathbf{D}_p$ is also equal to Hom(G, $\mathbf{Z}_p$).

Because of its importance, Hom(G, $\mathbf{T}$) has a special name, $\hat{G}$. We shall see (in Section 4) that $(\hat{G})\hat{} = G$, that is

$$\mathrm{Hom}(\mathrm{Hom}(G, \mathbf{T}), \mathbf{T}) = G.$$

(We point out that $\hat{G}$ is locally compact.)

2. HAAR MEASURE

Let μ denote Lebesgue measure on $G = \mathbf{R}^n$. It is obvious that μ has the following properties:

$$\text{If } U \text{ is a non-empty open set, then } \mu(U) > 0. \tag{2.1}$$

$$\text{If } C \text{ is a compact set, then } \mu(C) < \infty. \tag{2.2}$$

$$\text{If } x \in G \text{ and } X \text{ is measurable, then } \mu(X + x) = \mu(X), \text{ where } X + x = \{y + x : y \in X\}. \tag{2.3}$$

The first assertion follows from the fact that U contains a non-empty, n-cube; the second from the fact that each compact set is contained in a large bounded n-cube; and the third from the definition of Lebesgue measure.

It is trivial that counting measure μ on $\mathbf{Z}$ also has properties (2.1)–(2.3) and it is not hard to see that if $[0, 2\pi)$ is identified with $\mathbf{T} = \mathbf{R}/2\pi\mathbf{Z}$, then Lebesgue measure restricted to $[0, 2\pi)$ has also properties (2.1)–(2.3). (Addition is modulo 2π.)

These measures have the further property of being *regular*: that is, for every measurable set X,

$$\mu(X) = \sup_{\substack{C \subseteq X \\ C \text{ compact}}} \mu(C) = \inf_{\substack{U \supseteq X \\ U \text{ open}}} \mu(U). \tag{2.4}$$

A weaker form of (2.4) is

$$\mu(V) = \sup_{\substack{C \subseteq V \\ C \text{ compact}}} \mu(C), \quad \text{and} \quad \mu(X) = \inf_{\substack{U \supseteq X \\ U \text{ open}}} \mu(U) \tag{2.5}$$

for all sets X (measurable) and V (open).

If G is any locally compact group, a positive measure μ on G which obeys (2.1)–(2.3) and (2.5) is called a *right Haar measure*. If (2.3) is replaced by "$\mu(x + X) = \mu(X)$", then μ is called a *left Haar measure*.

We should specify, of course, the σ-algebra of sets on which μ is defined. We let $\Sigma = \Sigma(G)$ be the σ-algebra of sets generated by the open subsets of G. Then Σ is called the ring of *Borel* sets. A *Borel measure* on G is a measure on G with domain Σ.

A non-negative Borel measure ν on G is called *regular* if (2.4) holds for all Borel sets X. A signed Borel measure ν is regular if, in a Hahn decomposition $\nu = \nu_1 - \nu_2$ of ν into the difference of two non-negative measures, both ν_1 and ν_2 are regular. A complex Borel measure is regular if its real and imaginary parts are both regular. The following asserts the existence of a Haar measure for all locally compact groups:

THEOREM 2.1: *Let G be a locally compact group and Σ the σ-ring of Borel sets. Then there exists a real-valued non-negative measure $\mu: \Sigma \to \mathbf{R} \cup \{\infty\}$ which obeys* (2.1)-(2.3) *and* (2.5). *If ν is any other measure (non-negative, defined on Σ) which obeys* (2.1)-(2.3) *and* (2.5) *then there exists $\alpha > 0$ such that $\nu = \alpha\mu$.*

Note that for abelian groups G, a left Haar measure is also a right Haar measure; there are some groups for which left Haar measures are not right Haar measures (see for example, the article by Professor Sally in this book, pages 224–256). A Haar measure μ is regular when G is σ-compact (when μ is σ-finite), and, more generally, every σ-finite non-negative measure absolutely continuous with respect to Haar measure is regular. Finally, note that for the classical groups, a choice of Haar measure is easy, but for a group like $\mathrm{Hom}(\mathbf{T}_d, \mathbf{T})$, the existence of Haar measure is far from obvious. Haar measure plays a fundamental role in the study of general LCAG's.

When G is a compact group, we shall assume the Haar measure on G is normalized so G has mass one; when G is discrete, we shall assume Haar measure is normalized so it assigns mass one to singletons. (And we shall usually avoid talking about Haar measure on finite groups.)

Once having found a Haar measure μ, we define $L^1(G)$ to be the Lebesgue space $L^1(G, \Sigma, \mu)$ consisting of all Borel measurable functions $f: G \to \mathbf{C}$ such that $\int |f| \, d\mu = \|f\| < \infty$. (As usual, we identify functions equal except on a null set, so $L^1(G)$ is really a collection of equivalence classes of functions.)

When $f, g \in L^1(G)$, we can define a new function $f*g$ called the *convolution* of f and g as follows:

$$f*g(x) = \int f(x-y)g(y)\,d\mu(y). \tag{2.6}$$

With this multiplication $L^1(G)$ is an algebra, and $\|f*g\| \leq \|f\|\,\|g\|$. (These assertions follow by taking limits of simple functions and using Fubini's theorem.)

An element f of $L^1(G)$ may also be thought of as a measure ν_f: the value assigned to a Borel set X is the number $\nu_f(X)$ given by

$$\nu_f(X) = \int_X f(x)\,d\mu(x),$$

and we shall often write $\nu \in L^1(G)$ when ν is a finite measure on G which is absolutely continuous with respect to Haar measure.

3. OTHER MEASURES ON G

There are other regular Borel measures on G besides those in $L^1(G)$. For example, the unit point mass δ_x at some fixed $x \in G$: for each Borel set X,

$$\delta_x(X) = \begin{cases} 1, \text{if } x \in X, \\ 0, \text{if } x \notin X. \end{cases} \tag{3.1}$$

When $G = \mathbf{T}^2$, we can construct the product measure $\delta_0 \times \mu_2$, where μ_2 is Haar (Lebesgue) measure on the second factor of $\mathbf{T}^2$. This, too, is a regular measure. In Section 7, we construct a less trivial measure.

Given two regular Borel measures ν and ω, we define their *convolution* $\nu*\omega$ by the formula (X ranges over all Borel sets)

$$(\nu*\omega)(X) = \int \nu(X-x)\,d\omega(x). \tag{3.2}$$

(One must, of course, verify that $x \to \nu(X - x)$ is a Borel measurable function, and then that (3.2) defines a regular Borel measure on G.) An easy computation shows $\delta_0 * \nu = \nu$ for all regular Borel measures ν on G, so δ_0 is a multiplicative identity.

When the norm $\|\nu\|$ of a regular Borel measure ν is defined by $\|\nu\| = \sup|\int f\, d\nu|$, then $\|\nu * \omega\| \leqslant \|\nu\|\,\|\omega\|$, where the supremum here is taken over all continuous functions $f\colon G \to \mathbf{C}$ with $|f(x)| \leqslant 1$ for all $x \in G$. Under this norm, the set $M(G)$ of all regular Borel measures on G is a Banach space. With convolution multiplication it is a commutative algebra, and is usually called the *measure algebra* of G. (A Banach space which is also an algebra, and for which the multiplication is continuous, is called a *Banach algebra* so $M(G)$ and $L^1(G)$ are both Banach algebras.) Of course, we include in $M(G)$ only those measures ν for which $\|\nu\| < \infty$.

The measure algebra $M(G)$ is also represented as the dual space of $C_0(G)$, where $C_0(G)$ is the set of continuous complex-valued function f on G such that for each $\epsilon > 0$ there exists a compact set $C \subseteq G$ with $|f(x)| < \epsilon$, for all $x \in G$, $x \notin C$. The space $C_0(G)$ is a Banach space (with norm $\|f\| = \sup_{x \in G}|f(x)|$), and every continuous linear functional $\varphi\colon C_0(G) \to \mathbf{C}$ is given by integration against a (unique) regular Borel measure $\nu \in M(G) : \varphi(f) = \int f(x)\, d\nu(x)$ for all $f \in C_0(G)$. (This last assertion is called the *Riesz Representation Theorem.*)

The reader will note that for measures $\nu = \nu_f \in L^1(G)$ we have two possible convolutions and two possible norms: we can compute $\nu_f * \nu_g$, or ν_{f*g}. Of course these turn out to be the same: $\|\nu_f\| = \|f\|$, and $\nu_{f*g} = \nu_f * \nu_g$ for all $f, g \in L^1(G)$. Furthermore, $L^1(G)$ is an ideal in $M(G)$; that is, if $\omega \in M(G)$, and $f \in L^1(G)$, then $\omega * \nu_f = \nu_g$ where $g(x) = \int f(x - y)\, d\omega(y) \in L^1(G)$.

4. MULTIPLICATIVE LINEAR FUNCTIONALS

Because $M(G)$ and $L^1(G)$ are Banach algebras, as well as Banach spaces, it is fruitful to study the *multiplicative* linear functionals on $M(G)$ and $L^1(G)$. A linear function $\varphi\colon M(G) \to \mathbf{C}$ (or $\varphi\colon L^1(G) \to \mathbf{C}$) is a *multiplicative linear functional* (m.l.f.) if $\varphi(\nu * \omega) = \varphi(\nu)\varphi(\omega)$ for all ν, $\omega \in M(G)$ (or ν, $\omega \in L^1(G)$). All m.l.f.s are continuous. The dual space of $L^1(G)$ is $L^\infty(G) = L^\infty(G, \Sigma, \mu)$, the space of essentially bounded functions, that is, a Borel measurable function $f: G \to \mathbf{C}$ belongs to $L^\infty(G)$ if and only if there exists $C > 0$ such that $\{x : |f(x)| > C\}$ has zero μ (Haar) measure. Therefore, every m.l.f.φ on $L^1(G)$ is representable by integration against a Borel function h on G such that for some $C \geqslant 0$, $\{x \in G: |h(x)| > C\}$ has zero Haar measure. For $G = \mathbf{R}$, a straightforward application of Fubini's theorem to $\int h(x) f * g(x)\, d\mu(x)$ shows that integration against $h(x) = e^{-ix}$ is a m.l.f. on $L^1(\mathbf{R})$. Note that $h(x+y) = h(x)h(y)$ and that h is a continuous group homomorphism of $\mathbf{R}$ to T. This is typical:

THEOREM 4.1: *Let G be a LCAG, and let φ be a m.l.f. on $L^1(G)$. Then there exists a continuous group homomorphism γ: $G \to T$ such that*

$$\varphi(f) = \int f(x)\overline{\gamma(x)}\, d\mu(x)$$

for all $f \in L^1(G)$. If φ is integration against such a γ, then φ is a m.l.f. If $\int f(x)\gamma(x)\, d\mu(x) = 0$ for all continuous group homomorphisms γ, then $f \equiv 0$.

Recall that we wrote $\hat{G} = \mathrm{Hom}(G, \mathbf{T})$ for the collection of all continuous group homomorphisms from G to $\mathbf{T}$ and gave the $\hat{G}$ compact-open topology. For $f \in L^1(G)$, we define the *Fourier transform* $\hat{f}$: $\hat{G} \to \mathbf{C}$ by

$$\hat{f}(\gamma) = \int f(x)\overline{\gamma(x)}\, d\mu(x). \tag{4.1}$$

We now give $\hat{G} = \mathrm{Hom}(G, \mathbf{T})$ a "new" topology, the weakest topology such that all Fourier transforms $\hat{f}$, of elements $f \in L^1(G)$ are continuous. It is easy to see that the (old) pointwise multiplication $(\gamma\rho)\ (x) = \gamma(x)\rho(x) \in \mathbf{T}$ of elements γ, $\rho \in \mathrm{Hom}(G, \mathbf{T}) \subseteq L^\infty(G)$ is continuous in this topology, and that $\hat{G} = \mathrm{Hom}(G, \mathbf{T})$ is a LCAG in this "new" topology. This "new" topology is actually the same as the old compact-open topology. Note that G determines both $L^1(G)$ and $\hat{G}$, and that $L^1(G)$, by what we have just said, determines $\hat{G}$ also. We may now state the fundamental *Pontryagin Duality Theorem*:

THEOREM 4.2: $(\hat{G})\hat{} = G$.

This duality theorem is not established at once from what has been said. Because of the duality theorem, we shall write (x, γ) or (γ, x) for the value of $\gamma \in \hat{G}$ at $x \in G$ (and the value of $x \in \hat{\hat{G}} = G$ at $\gamma \in \hat{G}$). Formula (4.1) then becomes

$$\hat{f}(\gamma) = \int f(x)(x, -\gamma)d\mu(x). \tag{4.2}$$

Before continuing on to two major results which lead to a proof of Theorem 4.2, we suggest that the reader recall formulae (1.1).

The first major step in a proof of the duality theorem will be a generalization of the Riesz-Fischer Theorem.

THEOREM 4.3: (Riesz-Fischer). *If* $f \in L^2(\mathbf{T})$, *then* $\hat{f} \in L^2(\mathbf{Z})$ *and* $\|\hat{f}\|_2 = \|f\|_2$. *If* $g \in L^2(\mathbf{Z})$, *there exists* $f \in L^2(\mathbf{T})$ *such that* $\hat{f} = g$.

Because $\mathbf{T}$ has finite Haar (Lebesgue) measure, $L^2(\mathbf{T}) \subseteq L^1(\mathbf{T})$; when G is not compact, $\hat{f}$ does not make immediate sense for all $f \in L^2(G)$, since $L^2(G) \not\subseteq L^1(G)$. But we do have this generalization of 4.3.

THEOREM 4.4: (Plancherel Theorem). *Let G be a* LCAG *with Haar measure μ. Then there is a choice of Haar measure $\hat{\mu}$ on $\hat{G}$ such that, for every $f \in L^1(G) \cap L^2(G)$, $\hat{f} \in L^2(\hat{G})$ and $(\int |f|^2 d\mu)^{1/2} = (\int |\hat{f}|^2 d\hat{\mu})^{1/2}$. Furthermore, $\{\hat{f}: f \in L^1(G) \cap L^2(G)\}$ is dense in $L^2(\hat{G})$.*

Thus, the Fourier transform can be extended to an isometric isomorphism of $L^2(G)$ onto $L^2(\hat{G})$.

The choice of Haar measure $\hat{\mu}$ in the Plancherel theorem, in addition to making $L^2(G)$ and $L^2(G)$ isometrically isomorphic via the (extended) Fourier transform, also has the property that if $f \in L^1(G)$ has $\hat{f} \in L^1(\hat{G})$, then

$$f(x) = \int \hat{f}(\gamma)(x, \gamma)d\hat{\mu}(\gamma). \tag{4.3}$$

This is called the *inversion formula*. The set of $f \in L^1(G)$ such that $\hat{f} \in L^1(\hat{G})$ is dense in $L^1(G)$ and the corresponding $\hat{f}$'s are dense in $L^1(\hat{G})$. Note that in (4.3), we integrate $\hat{f}$ against (x, γ) and not against $(x, -\gamma)$, as in (4.2).

The second step toward a proof of theorem 4.1 is a generalization of Herglotz's Theorem concerning positive-definite functions on **Z**. A complex valued function φ: $G \to \mathbf{C}$ on a LCAG is *positive-definite* if $n \geqslant 1$, $x_1 \in G, \ldots, x_n \in G$, $c_1 \in \mathbf{C}, \ldots, c_n \in \mathbf{C}$ imply

$$\sum_{j,k=1}^{n} c_j \bar{c}_k \varphi(x_j - x_k) \geqslant 0.$$

Let ν be a regular Borel measure on G. The *Fourier-Stieltjes transform* of ν is the function

$$\hat{\nu}(\gamma) = \int_G (x, -\gamma)d\nu(x), \tag{4.4}$$

from $\hat{G}$ to **C**. Each $\hat{\nu}$ is continuous on $\hat{G}$, and if ν is real and non-negative, then $\hat{\nu}$ is positive-definite on $\hat{G}$ because

$$\sum_{j,k} c_j \bar{c}_k \hat{\nu}(\gamma_j - \gamma_k) = \int \int \sum c_j \bar{c}_k (x, \gamma_k - \gamma_j) d\nu(x)$$

$$= \int |\sum c_j (x, \gamma_j)|^2 d\nu \geqslant 0.$$

Herglotz's Theorem says that the only positive-definite functions on **Z** are the Fourier-Stieltjes transforms of non-negative measures on **T**. *Bochner's Theorem* is the generalization of this to other LCAG's:

THEOREM 4.5: *Let φ be a continuous positive-definite function on a* LCAG G. *Then there exists a real non-negative regular Borel measure ν on $\hat{G}$ such that φ is the Fourier-Stieltjes transform of ν:*

$$\varphi(x) = \int_{\hat{G}} (-x, \gamma) d\nu(\gamma), \; x \in G.$$

We conclude this section with some facts about the Fourier-Stieltjes transform. First, if $\nu_f \in L^1(G)$, then $\hat{\nu}_f(\gamma) = \hat{f}(\gamma)$. Secondly, if $\hat{\nu}(\gamma) = 0$ for all $\gamma \in \hat{G}$, then $\nu = 0$. The Fourier-Stieltjes transform is multiplicative: $(\omega * \nu)\hat{}(\gamma) = \hat{\omega}(\gamma)\hat{\nu}(\gamma)$. (Not all m.l.f.s on $M(G)$ are given by $\nu \to \hat{\nu}(\gamma)$, $\gamma \in \hat{G}$, however.) Finally, if $\nu \in L^1(G)$, then $\hat{\nu} \in C_0(\hat{G})$. ($C_0(\hat{G})$ is defined in Section 3.) This last result is called the *Riemann-Lebesgue Lemma*. When G is not discrete, there do exist measures $\nu \in M(G)$, $\nu \not\in L^1(G)$ for which $\hat{\nu} \in C_0(\hat{G})$; if $x \in G$, $\hat{\delta}_x(\gamma) = (x, -\gamma)$ has constant modulus one, so not all $\nu \in M(G)$ have $\hat{\nu} \in C_0(G)$.

In the preceding paragraph, we asserted that not all m.l.f.s on $M(G)$ were given by evaluation of the Fourier-Stieltjes transform at an element of G. Here is an example. A regular Borel measure $\nu \in M(G)$ is *continuous* if $\nu(\{x\}) = 0$ for all $x \in G$, and ν is *discrete* if ν is an (absolutely convergent) weighted sum, $\Sigma a_j \delta_{x_j}$, of point masses. Every $\nu \in M(G)$ has a unique expression as a sum $\nu = \nu_d + \nu_c$ of a discrete measure ν_d and a continuous measure ν_c. We define a m.l.f. φ: $M(G) \to \mathbf{C}$ by $\varphi(\nu) = (\nu_d)\hat{}\,(0)$ where 0 is the identity of $\hat{G}$. That φ is indeed a m.l.f. follows from the fact that the set of continuous (regular Borel) measures on G is a closed ideal in $M(G)$.

We now discuss homomorphisms between LCAG's and their associated algebras.

When G and H are LCAGs and $\varphi \in \text{Hom}(G, H)$ is a continuous group homomorphism from G to H, a map $\check{\varphi}$: $M(G) \to M(H)$ is defined as follows:

$$(\check{\varphi}\nu)(X) = \nu\big(\varphi^{-1}(X)\big), \; X \in \Sigma(H). \tag{4.5}$$

Then $\check{\varphi}$ is a continuous algebra homomorphism with $\|\check{\varphi}\nu\| \leqslant \|\nu\|$.

We can also define $\hat{\varphi} \in \mathrm{Hom}(\hat{H}, \hat{G})$ by

$$[\hat{\varphi}(\gamma)](x) = \gamma(\varphi(x)). \tag{4.6}$$

(It is not hard to show, by computation, that (4.5) and (4.6) define maps with the desired properties; for example, that $\hat{\varphi}$ is a continuous group homomorphism); also, $\hat{\varphi}$ is one-to-one if φ is onto. Here are three examples:

(i) Let $\varphi: \mathbf{R} \to T$ be the map $x \to e^{ix}$. Then $\hat{\varphi}(\mathbf{Z}) = \mathbf{Z} \subseteq \mathbf{R}$, and if $\nu \in M(\mathbf{R})$, then $(\check{\varphi}\nu)\hat{}(n) = \hat{\nu}(n)$, and $\check{\varphi}\nu(X) = \nu(\cup X'^{\infty}_{k=-\infty} + 2\pi k)$ where $X' = \{x \in [0, 2\pi): e^{ix} \in X\}$ for all Borel sets $X \subseteq \mathbf{R}$.

(ii) Let $\psi: \mathbf{Z} \to \mathbf{R}$ be given by $\psi(n) = n$ (so $\psi = \hat{\varphi}$, where φ is as in (i) and $\hat{\psi} = \varphi$). Let $\nu \in M(\mathbf{Z}) = L^1(\mathbf{Z})$. Then $(\check{\psi}\nu)\hat{}(x) = \hat{\nu}(\hat{\psi}(x)) = \hat{\nu}(\hat{\psi}(x + 2\pi))$. This tells us that $(\check{\psi}\nu)\hat{}$ is 2π-periodic on $\mathbf{R}$. Conversely if $\omega \in M(\mathbf{R})$ and $\hat{\omega}$ is 2π-periodic, then ω is a sum $\Sigma a_j \delta_{m_j}$, $m_j \in \mathbf{Z}$, $\Sigma|a_j| < \infty$.

(iii) Let $G = \mathbf{R}$, $H = \mathbf{T}^2$, and let $\varphi(x) = (e^{ix}, e^{i\pi\sqrt{2}\,x})$. Then φ is a continuous group homomorphism from $\mathbf{R}$ to $\mathbf{T}^2$. Because $\sqrt{2}$ is irrational, φ is actually one-to-one, though not onto, and $\varphi(\mathbf{R})$ is dense in $\mathbf{T}^2$. The map $\hat{\varphi}$ sends $(m, n) \in \mathbf{Z}^2 = (\mathbf{T}^2)\hat{}$ to $m(n/\sqrt{2})$. We let $\varphi(\mathbf{R})$ be an open (hence closed) subgroup of H, whose topology is exactly the same as $\mathbf{R}$'s, that is, φ is now a homeomorphism from $\mathbf{R}$ to $\varphi(\mathbf{R}) \subseteq H$. (This is the same group (for $\xi = \sqrt{2}$) which was defined at the end of Section 1.)

5. THE STRUCTURE THEOREM

One way to construct new LCAG's is to make product groups; for example, one can take $G = \mathbf{R} \times \mathbf{T} \times \mathbf{Z}$, with the product topology, and coordinate-wise group operations. A tedious but not particularly difficult argument shows that G is then a LCAG. Note that $\mathbf{R} \times \mathbf{T} \times \{0\}$ is an open subgroup of G.

The structure theorem tells us that $\mathbf{R} \times \mathbf{T} \times \mathbf{Z}$ is more or less typical. By $\mathbf{R}^0$ we mean the group with one element.

THE STRUCTURE THEOREM 5.1: *Let G be a* LCAG. *Then there exists an integer $n \geqslant 0$ such that*

(i) *G contains an open subgroup H which is isomorphic to a product $\mathbf{R}^n \times C$, where C is a compact abelian group.*

(ii) *G is isomorphic to a product $\mathbf{R}^n \times K$, where K is a* LCAG *with a compact open subgroup C'.*

It is worth observing that while n is determined by G, the compact subgroups C and C' are not. The cyclic group $\mathbf{Z}_4$ of order four provides the necessary example: the candidates for C and C' are $\{0\}$, $\{0, 2\}$, and $\mathbf{Z}_4$ itself.

If H is the open subgroup of (i), then G/H is discrete and one might hope that G would be isomorphic to a product of $\mathbf{R}^n$, a compact group and a discrete group. Unfortunately there are examples which show that this is not the case.

6. THE IDEMPOTENT THEOREM

In the study of the measure algebra $M(G)$, it is very useful to know which regular Borel measures $\nu \in M(G)$ are idempotent (that is, $\nu * \nu = \nu$). Before stating the theorem, we will consider some examples.

Let G be a LCA group and suppose H is a *compact* subgroup of G. Then H is a LCAG, and so that Haar measure, ν, of H which has $\nu(H) = 1$, is an element of $M(G)$. Also (3.2) and the translation invariance of ν show that $\nu * \nu = \nu$. The measure $\delta_0 \times \mu_2$ on $\mathbf{T}^2$ which we mentioned in 3.1 is an example of such a ν. We will call such a measure "Haar measure of a compact subgroup." Note that δ_0 is one of these examples also, since $\{0\}$ is a compact subgroup.

If ν is idempotent, then $\delta_0 - \nu$ is idempotent since $(\delta_0 - \nu) * (\delta_0 - \nu) = \delta_0^2 - 2\delta_0 * \nu + \nu^2 = \delta_0 - 2\nu + \nu = \delta_0 - \nu$. (Recall that δ_0 is the identity of $M(G)$.) We have thus obtained a new idempotent from ν by what we will call the operation of complementation.

If ν and ω are both idempotent, then $\nu * \omega$ is idempotent. (The operation of intersection.) Furthermore, $\nu + \omega - \nu * \omega$ is idempotent (operation of union). These assertions can be verified by straightforward computations.

There is one further method for obtaining new idempotents from old. Because a measure ν is determined by its Fourier-Stieltjes transform, $\hat{\nu}$ is idempotent, if and only if $\hat{\nu}$ maps $\hat{G}$ to $\{0, 1\}$. Given a measure ν, and $\gamma \in \hat{G}$ we define a new measure $\gamma\nu$, given by

$$\int f(x)d(\gamma\nu)(x) = \int f(x)(x, \gamma)d\nu(x),$$

where f is any bounded Borel function. A simple computation shows that the Fourier-Stieltjes transform of $\gamma\nu$ is just the "translation" of the Fourier-Stieltjes transform of ν:

$$(\gamma\nu)\hat{}(\rho) = \hat{\nu}(\rho - \gamma), \quad \rho \in \hat{G}. \tag{6.1}$$

If ν is idempotent, then $\gamma\nu$ is also idempotent, since the range of $(\gamma\nu)\hat{}$ on $\hat{G}$ is, by (6.1), exactly the range of $\hat{\nu}$. We will say $\gamma\nu$ is obtained from ν by the operation of translation.

We can now state our theorem.

THEOREM 6.1: *Let G be a* LCAG *and $\nu \in M(G)$ an idempotent measure. Then there exist a finite number of Haar measures $\nu_1, \ldots, \nu_n$ of compact subgroups of G such that ν can be obtained from $\nu_1, \ldots, \nu_n$ by a finite number of applications of the operations of complementation, intersection, union, and translation.*

This theorem is usually called the *Cohen Idempotent Theorem*, because it was proved for general LCAG's by Paul Cohen. Henry Helson proved this theorem when $G = \mathbf{T}$, and Walter Rudin proved it for $G = \mathbf{T}^n$.

The reader probably found the terms, "complementation", "intersection", "product", and "translation" peculiar. We will now explain them, and give another view of the idempotent theorem.

If $\nu \in M(G)$ is idempotent, let $E_\nu = \{\gamma \in \hat{G}: \hat{\nu}(\gamma) = 1\}$. Then

E_ν is an open and closed set. If ν is Haar measure on a compact subgroup H of G, then

$$E_\nu = \{\gamma \in \hat{G} : (\gamma, x) = 1 \text{ for } x \in H\}.$$

It is easy to see that E_ν is an open and closed subgroup of $\hat{G}$. Also, it is not hard to show that if E is an open and closed subgroup of $\hat{G}$, then there exists a Haar measure ν on a compact subgroup of G such that $E = E_\nu$.

If $\gamma \in \hat{G}$, then $E_{\gamma\nu} = \{\rho - \gamma : \rho \in E_{\gamma\nu} = E_\nu - \gamma\}$ by (6.1), and $E_{\gamma\nu}$ is an open and closed coset in G. The *coset ring* of $\hat{G}$ is the set of all subsets of $\hat{G}$ which can be generated from the open and closed cosets of subgroups of $\hat{G}$ in a finite number of steps using the operations of translation (replacing E by $E + \gamma$), complementation (replacing E by $\{\gamma \in \hat{G}: \gamma \not\subseteq E\}$), union and intersection. Straightforward computations show that these operations on sets of the form E_ν correspond to the same named operations on idempotent measures. The Cohen Idempotent Theorem then has the following form:

THEOREM 6.2: *Let $E \subseteq \hat{G}$. Then the characteristic function of E is a Fourier-Stieltjes transform if and only if E belongs to the coset ring.*

(The "if" part of this follows trivially.) It is in this form that Helson, Rudin and Cohen proved Theorem 6.1.

Finally, let us consider what this says when $G = T$. Then the property "E belongs to the coset ring" is the same as "outside of a finite set, E is periodic." This is because the only sub-groups of $\mathbf{Z} = \mathbf{T}^\wedge$ are the sets of the form

$$\{kn : n \in \mathbf{Z}\}, k = 0, 1, 2, \ldots.$$

A few moments' thought should then convince the reader that the statements in quotes are equivalent.

Some sets can be shown not to be in the coset ring without using Theorem 6.2. For example, let G be compact and let $E \subseteq \hat{G}$ be an infinite set such that each element $\gamma \in \hat{G}$ has *at most* one expression as a sum

$$\gamma = \pm\gamma_1 \pm \cdots \pm \gamma_n \tag{6.2}$$

where $\gamma_1, \ldots, \gamma_n$ are distinct elements of E. (Such a set is *dissociate.*) We now show how one proves that E is not in the coset ring. Fix an integer $n \geqslant 1$ and let $\gamma_1, \ldots, \gamma_{2n} \in E$ be distinct.

Define the trigonometric polynomial $P(x)$ by

$$P(x) = \prod_{j=1}^{2n} \left(1 + \frac{i}{\sqrt{n}} \left(\tfrac{1}{2}(x, \gamma_j) + \tfrac{1}{2}(x, -\gamma_j)\right)\right).$$

A straightforward calculation shows that

$$|P(x)| \leqslant \prod_{1}^{2n} |1 + i/\sqrt{n}\,| = (1 + 1/n)^n \leqslant e.$$

Slightly more complex calculations (using the dissociate property (6.2) of E) show that the Fourier coefficient of P at $\gamma_j (1 \leqslant j \leqslant 2n)$ is $i/(2\sqrt{n})$, and zero for all other γ's in E. Thus, if $\nu \in M(G)$ has $\hat{\nu} = 1$ on E and $\hat{\nu} = 0$ off E, then $\int P(d\nu) = \Sigma \hat{P}(\gamma)\hat{\nu}(\gamma) = 2in/2\sqrt{n} = i\sqrt{n}$. Thus, $\|\nu\| \geqslant \sqrt{n}/e$. Since n is arbitrary, this shows $\hat{\nu}$ cannot be the characteristic function of E, if ν is a measure with finite mass on G.

7. INVERSES IN $M(G)$

The idempotent theorem characterizes idempotent measures ν in the measure algebra $M(G)$ of a LCAG G, in terms of the behavior of the Fourier-Stieltjes transform $\hat{\nu}$ on $\hat{G}$. In studying an algebra with identity, one often wants to know which elements are "invertible." A measure $\nu \in M(G)$ is *invertible* if and only if there exists $\omega \in M(G)$ such that

$$\nu * \omega = \delta_0. \tag{7.1}$$

Considering Fourier-Stieltjes transforms, we see that (7.1) holds if and only if

$$\hat{\nu}(\gamma)\hat{\omega}(\gamma) = 1, \quad \text{all } \gamma \in \hat{G}. \tag{7.2}$$

Because $|\hat{\omega}(\gamma)| \leqslant \|\omega\|$, a *necessary* condition for (7.2) is that $\hat{\nu}(\gamma)$ be bounded away from zero; that is that

$$\inf_{\gamma \in \hat{G}} |\hat{\nu}(\gamma)| > 0. \tag{7.3}$$

A natural conjecture is that (7.3) is also *sufficient* for ν to be invertible. This is *not* the case, as the following theorem shows:

THEOREM 7.1: *Let G be a non-discrete* LCAG. *Then there exists a regular Borel measure $\nu \in M(G)$ such that*
(i) *the Fourier-Stieltjes transform of ν obeys* (7.3), *and*
(ii) *ν is not invertible in $M(G)$.*

This was stated by Norbert Wiener and R. H. Pitt in 1938 for $G = \mathbf{R}$. Their proof was incorrect; a correct proof (for $G = \mathbf{R}$) was given by Y. A. Sreider in 1948, and proofs for more general groups have been provided by J. H. Williamson (1958), E. Hewitt and S. Kakutani (also 1958), and others. Theorem 7.1 is usually called "The Wiener-Pitt Theorem."

If G is discrete, so $\hat{G}$ is compact, then $M(G) = L^1(G)$, and (7.3) *is* equivalent to the invertibility of ν. If G is not discrete and $\nu \in L^1(G)$, then the Riemann-Lebesgue Lemma (Section 4) shows $\hat{\nu} \in C_0(\hat{G})$ so $\inf|\hat{\nu}(\gamma)| = 0$, and ν is not invertible. With more effort, one can show that if ν is a sum of a measure in $L^1(G)$ and a discrete measure (a sum of point masses) and (7.3) holds, then ν *is* invertible. Therefore, a "bad" measure will be far from being in $L^1(G)$.

We give here a proof of Theorem 7.1 for $G = \mathbf{T}$. This proof may be extended, via the structure theorem, to all LCA groups and is typical in its use of functional analysis. We begin with some preliminary remarks.

The *spectrum* of $\nu \in M(\mathbf{T})$ is the set of complex numbers λ such that $\nu - \lambda\delta_0$ is *not* invertible. The spectrum $\sigma(\nu)$ of ν has the following properties:

(i) $\sigma(\nu) \neq \emptyset$;
(ii) $\sup_{\lambda\in\sigma(\nu)}|\lambda| = \lim_{n\to\infty}\sup\|\nu^n\|^{1/n} = \lim_{n\to\infty}\inf\|\nu^n\|^{1/n}$;
(iii) $\sigma(\nu) = \{\varphi(\nu) : \varphi : M(\mathbf{T}) \to \mathbf{C} \text{ is a m.l.f.}\}$;
(iv) if $X \subseteq \sigma(\nu)$ is a (relatively) open and closed subset, then there exists an idempotent measure ω such that if φ: $M(T) \to \mathbf{C}$ is any m.l.f., then $\varphi(\omega) = 1$ if $\varphi(\nu) \in X$ and $\varphi(\omega) = 0$ if $\varphi(\nu) \notin X$; finally
(v) $\hat{\nu}(\mathbf{Z}) \subseteq \sigma(\nu)$

Note that (v) follows from (iii) and the fact (Section 4) that $\nu \to \hat{\nu}(\gamma)$, $\gamma \in \mathbf{Z} = \mathbf{T}\hat{}$ is multiplicative, and that (iv) follows from the Šilov Idempotent Theorem (see [4], [12], or [14]).

If $\nu \in M(\mathbf{T})$ is such that $\hat{\nu}(\mathbf{Z})$ is not dense in $\sigma(\nu)$, then for some $\lambda \in \sigma(\nu)$,

$$\inf_{\gamma\in\mathbf{Z}}|\hat{\nu}(\gamma) - \lambda| = \inf_{\gamma\in\mathbf{Z}}|(\nu - \lambda\delta_0)\hat{}(\gamma)| > 0,$$

while $\nu - \lambda\delta_0$ is not invertible. Thus, to prove Theorem 7.1, it will be sufficient to find $\nu \in M(\mathbf{T})$ such that $\hat{\nu}(\mathbf{Z})$ is not dense in $\sigma(\nu)$. We first fix an integer $n \geqslant 1$, and let

$$P_n(x) = \prod_{j=1}^{n}(1 + \cos 10^j x)$$

$$= \prod_{j=1}^{n}\left(1 + \tfrac{1}{2}e^{i10^j x} + \tfrac{1}{2}e^{-i10^j x}\right). \tag{7.4}$$

It is easy to see that

$$P_n\hat{}(\gamma) = \begin{cases} 1, & \text{if } \gamma = 0, \\ 2^{-k}, & \text{if } \gamma = \pm 10^{j_1} \pm \cdots \pm 10^{j_k}, \\ & \quad 1 \leqslant j_1 < \cdots < j_k \leqslant n, \\ 0, & \text{otherwise.} \end{cases} \tag{7.5}$$

Since each factor in (7.5) is non-negative, $\|P_n\| = \int P_n(x)\,dx = 1$, so the measure corresponding to $P_n \in L^1(\mathbf{T})$ is a positive measure of norm one. We now let $n \to \infty$. If f is any continuous function on $\mathbf{T}$, then f is the uniform limit of trigonometric polynomials, so for any $\epsilon > 0$, there exists a trigonometric polynomial Q such that $\sup_{x\in\mathbf{T}}|f(x) - Q(x)| < \epsilon$. Then

$$\left|\int f(x)P_n(x)\,dx - \int Q(x)P_n(x)\,dx\right| < \epsilon, \quad \text{for } n = 1, 2, \ldots . \tag{7.6}$$

From (7.5), it follows that

$$\lim_{n\to\infty}\int Q(x)P_n(x)\,dx \tag{7.7}$$

exists for every trigonometric polynomial $Q(x) = \sum_{k=1}^{l} a_k e^{im_k x}$. Then, (7.6) and (7.7) together imply that

$$\lim_{n\to\infty}\int f(x)P_n(x)\,dx$$

exists for each $f \in C(\mathbf{T})$ and defines a continuous linear functional on $C(\mathbf{T})$. The Riesz representation theorem tells us that there exists a unique measure $\nu \in M(\mathbf{T})$ such that

$$\int f(x)d\nu(x) = \lim_{n\to\infty}\int f(x)Pn(x)\,dx.$$

Putting $e^{-i\gamma x}$ for f, we see

$$\hat{\nu}(\gamma) = \begin{cases} 1, & \text{if } \gamma = 0, \\ 2^{-k}, & \text{if } \gamma = \pm 10^{j_1} \cdots \pm 10^{j_k} \\ & \quad 1 \leqslant j_1 < \cdots < j_k < \infty, \\ 0, & \text{otherwise.} \end{cases} \tag{7.8}$$

Thus, $\hat{\nu}(\mathbf{Z}) = \{0, 1, 1/2, 1/4, 1/8, \ldots\}$.

We now show that $\frac{1}{2}$ is not isolated in $\sigma(\nu)$; that is, there exists

$\lambda \in \sigma(\nu)$ such that

$$0 < |\lambda - 1/2| < 1/8.$$

This will then show that $\hat{\nu}(\mathbf{Z})$ is not dense in $\sigma(\nu)$ and thus complete the proof.

So suppose $\frac{1}{2}$ were isolated in $\sigma(\nu)$. Then by (iv), there would exist $\omega \in M(\mathbf{T})$ such that $\hat{\omega}(\gamma) = 1$ if $\hat{\nu}(\gamma) = 1/2$ and $\hat{\omega}(\gamma) = 0$ if $\hat{\nu}(\gamma) \neq 1/2$. Then ω is an idempotent measure so (by Theorem 6.2), $\{\gamma : \hat{\omega}(\gamma) = 1\}$ is periodic except for a finite set. This is absurd since

$$\begin{aligned} \sum &= \{10^j : 1 \leqslant j < \infty\} \cup \{-10^j : 1 \leqslant j < \infty\} \\ &= \{\gamma : \hat{\omega}(\gamma) = 1\} \end{aligned}$$

is not periodic. Thus $\frac{1}{2}$ is not isolated in $\sigma(\nu)$.

REMARKS. The method used at the end of Section 6 can be modified to show that E is not of the form $\{\gamma : \hat{\omega}(\gamma) = 1\}$ for an idempotent $\omega \in M(\mathbf{T})$; as at the end of Section 6, this method will avoid using Theorem 6.2. The measure ν used above and the polynomials in Section 6 are called "Riesz Products"; a detailed discussion appears in vol. I of [18].

PART TWO. SOME SPECIFIC PROBLEMS. SECTIONS 8–13.

The ideal theory for $L^1(G)$ is discussed in Sections 8 and 9. In §10 we discuss a notion of "thinness" for subsets E of a LCAG G, and in §11, relate it to both the ideal theory, and to convergence of (formal) Fourier series. In §12 we discuss Cantor sets (like the middle-third set) and indicate some relations between number theory and harmonic analysis properties of Cantor Sets. In §13 we give L. Schwartz's example of a set of "non-synthesis" in $\mathbf{R}^3$ (this fulfills a promise made in §8). Throughout this second part we will try to point out how non-classical groups intervene in the proofs of theorems about classical groups.

8. IDEALS IN $L^1(G)$

A natural question to ask about an algebra is: What are its ideals? When the algebra has a topology, we may ask, what are its closed ideals?

For $L^1(G)$, we first observe that if $\gamma \in \hat{G}$, then $\{f \in L^1(G) : \hat{f}(\gamma) = 0\}$ is a closed ideal, because the map $f \to \hat{f}(\gamma)$ is a *multiplicative* (continuous) linear functional on $L^1(G)$. (That is how we obtained $\hat{G}$ in Section 4). Of course, the intersection of a set of closed ideals is again a closed ideal. Therefore, if $Y \subseteq \hat{G}$, then

$$I(Y) = \{f \in L^1(G) : \hat{f}(\gamma) = 0, \text{all } \gamma \in Y\} \qquad (8.1)$$

is a closed ideal. Since each $\hat{f}$ (for $f \in L^1(G)$) is continuous on $\hat{G}$, $I(Y) = I(Y^-)$ where $^-$ denotes closure in G. The ideal $I(Y)$ is called the kernel of Y. Of course, it is *a priori* possible that $I(Y) = \{0\}$. The part (i) of the following Theorem shows (in particular) that $I(Y) \neq 0$ if $Y^- \neq \hat{G}$. The notation used is this: if K is any ideal of $L^1(G)$, $Z(K) = \{\gamma \in \hat{G} : \hat{f}(\gamma) = 0, \text{ all } f \in K\}$. (The set $Z(K)$ is called the hull of the ideal K.)

THEOREM 8.1: *Let G be a* LCAG. *Let $Y \subseteq \hat{G}$ be a closed set and $K \subseteq L^1(G)$ a closed ideal. Then*

(i) *if $C \subseteq \hat{G}$ is a compact set and $C \cap Y = \varnothing$, there exists $f \in L^1(G)$ with $\hat{f}(\gamma) = 0$, all $\gamma \in Y$ and $\hat{f}(\gamma) = 1$, all $\gamma \in C$.*

(ii) $Z(I(Y)) = Y$.

(iii) *If $Z(K) = \varnothing$, then $K = L^1(G)$.*

(iv) *$I(Z(K)) \supseteq K$.*

The proof of (i) is somewhat difficult; the other assertions follow easily from (i) and the definitions of $I(Y)$ and $Z(K)$.

Statement (i) suggests the following definition of an ideal which is (possibly) smaller than $I(Y)$. Let $J_0(Y)$ denote the set of $f \in L^1(G)$ such that $\hat{f}(\gamma) = 0$ for all γ in some neighborhood of Y. (The neighborhood varies with f.) $J_0(Y)$ may not be closed in the norm topology of $L^1(G)$, so we let $J(Y)$ be the closure of $J_0(Y)$. It is obvious that

$$J(Y) \subseteq I(Y)$$

and Theorem 8.1 (i) shows (easily) that

$$Z(J(Y)) = Y.$$

A more complicated argument shows that, with the hypotheses of Theorem 8.1:

COROLLARY 8.2: $J(Z(K)) \subseteq K$.

In words, $J(Y)$ is the smallest closed ideal K with $Z(K) = Y$ and $I(Y)$ is the largest ideal K with $Z(K) = Y$. We thus may ask, does $Y = Z(K)$ determine the closed ideal? That is, must $I(Y) = J(Y)$?

This is not obvious. The easiest example is found by looking at another Banach algebra: the Banach space $C^1(0, 1)$ consisting of all complex-valued continuous functions on $[0, 1]$ with continuous (first) derivatives. For $f \in C^1(0, 1)$, we define the norm of f by

$$\|f\| = \sup_{0 \leqslant x \leqslant 1} |f(x)| + \sup_{0 \leqslant y \leqslant 1} |f'(y)|. \tag{8.2}$$

It is not hard to see (compute) that with the norm (8.2), $C^1(0, 1)$ is a Banach space (with pointwise addition and multiplication of functions), an algebra, and $\|fg\| \leqslant \|f\| \cdot \|g\|$. It is somewhat harder to show that the maps $f \to f(x)(0 \leqslant x \leqslant 1)$ are *all* the multiplicative (continuous) linear functionals on $C^1(0, 1)$. We make the analogous definitions of $I(Y)$ and $Z(K)$, and consider the singleton set $\{\frac{1}{2}\}$. Let I be the closed ideal of $f \in C^1(0, 1)$ such that $f(\frac{1}{2}) = 0$; let J_0 be the ideal consisting of $f \in C^1(0, 1)$ which vanish in a neighborhood of $\frac{1}{2}$; and let J be the closure (in $C^1(0, 1)$) of J_0. Then every element $f \in J_0$ has $f'(\frac{1}{2}) = 0$, so, on applying (8.2), we see that if $f \in J$, $f'(\frac{1}{2}) = 0$ also. On the other hand, the function

$$f(x) = x - \tfrac{1}{2} (0 \leqslant x \leqslant 1.)$$

is in $C^1(0, 1)$, belongs to I, and not to J, since $f'(\frac{1}{2}) = 1$. Thus I and J are two closed ideals of $C^1(0, 1)$ with $Z(I) = Z(J)$ and $I \neq J$. We now return to $L^1(G)$ and make the following definition:

DEFINITION 8.3: A closed subset $Y \subseteq \hat{G}$ is a set of *spectral synthesis* (for $L^1(G)$) if $I(Y) = J(Y)$.

EXAMPLES: The following sets have been shown to be sets of spectral synthesis:

(i) Finite sets, closed countable sets.
(ii) For $G = \mathbf{R}$, any finite union of closed intervals.
(iii) For $G = \mathbf{R}^n$, any compact "star-shaped" set C—that is, there is a point x in the interior of C such that each line through x (infinite in both directions) meets the boundary C in exactly two points. (For example, the n-cube.)
(iv) Any set in $\mathbf{R}^2$ whose boundary is a finite union of finite line segments.
(v) Cantor's "middle third" set (see Section 12 below.)
(vi) All of $\hat{G}$.

The question of determining sets of spectral synthesis (we will explain the origin of the term below) was raised in the 1930's by A. Beurling. In 1948, Laurent Schwartz proved (recall that $\mathbf{R}^{3\wedge} = \mathbf{R}^3$):

THEOREM 8.4: *The unit sphere S in $\mathbf{R}^3$ is not a set of spectral synthesis for $L^1(\mathbf{R}^3)$.*

Schwartz's proof uses distributions and is given in Section 13 below. His method does not (immediately, at least) produce a set of non-synthesis for $L^1(\mathbf{R})$. This was done by Paul Malliavin in 1959:

THEOREM 8.5: *Let G be any non-compact* LCAG. *Then there exists a compact subset $Y \subseteq \hat{G}$ which is <u>not</u> a set of spectral synthesis.*

(If G is compact, $\hat{G}$ is discrete, so any compact subset is finite and hence of spectral synthesis, by (i) above.)

Malliavin's proof, for $G = \mathbf{R}$, does not use non-classical groups. In 1966, N. Th. Varopoulos gave a new proof of Malliavin's Theorem. This proof used the example of Schwartz, and then by ingenious techniques passed first to $\mathbf{D}_2$, and then to $\mathbf{R}$ (or any other non-discrete LCAG).

In the next section we describe another way of looking at spectral synthesis.

9. A REFORMULATION OF "SPECTRAL SYNTHESIS"

Let Y be a closed subset of G, where G is a LCAG. Note that $I(Y)$ and $J(Y)$ are closed *subspaces* of $L^1(G)$. If $I(Y) \neq J(Y)$, then an easy application of the Hann-Banach theorem shows that there exists a continuous linear functional φ: $L^1(G) \to \mathbf{C}$ such that $\varphi(f) = 0$ for all $f \in J(Y)$, and, for some $g \in I(Y)$, $\varphi(g) \neq 0$. Of course, since $L^\infty(G)$ is the dual space of $L^1(G)$, φ is represented by integration against an element $h \in L^\infty(G)$:

$$\varphi(f) = \int h(x) f(x)\, dx, f \in L^1(G), \tag{9.1}$$

where "dx" denotes integration with respect to Haar measure.

A bit further along we shall learn more about h. First, we consider some candidates for h. For example, let ν be a regular Borel measure on $\hat{G}$ (note the "hat"), and let $\hat{\nu}$ denote the Fourier-Stieltjes transform of ν. If $f \in L^1(G)$, then an application of Fubini's Theorem shows

$$\int_{\hat{G}} \hat{f}(\gamma) d\nu(\gamma) = \int_G \hat{\nu}(x) f(x)\, dx. \tag{9.2}$$

Of course, $\hat{\nu} \in L^\infty(G)$, since $|\hat{\nu}| \leqslant \|\nu\|$ and $\hat{\nu}$ is continuous on G.

For a measure ν on $\hat{G}$ (note the hat, again) we have the notion of "support": The *support* of ν is the smallest *closed* set $Y \subseteq \hat{G}$ such that if k is a continuous function on $\hat{G}$ which is zero on a neighborhood of Y, then $\int k(\gamma) d\nu(\gamma) = 0$. The support of ν is unchanged, if we replace "$k \in C(\hat{G})$, k zero on a neighborhood of Y" with "$k = \hat{f}, f \in L^1(G)$, $\hat{f}$ zero on a neighborhood of Y."

(This uses the easily proved density of $\{\hat{f} : f \in L^1(G)\}$ in $C_0(G)$ and 8.1).

With this change, we find:

The support of ν is the smallest closed set Y such that $\int \hat{f} d\nu = 0$ for all $f \in J(Y)$. (Of course, $\int \hat{g} d\nu = 0$ if $g \in I(Y)$ also.)

We now define the spectrum of an element $h \in L^\infty(G)$:

DEFINITION 9.1: The *spectrum* of $h \in L^\infty(G)$ is the smallest closed set $Y \subseteq \hat{G}$ such that $\int f(x)h(x)\,dx = 0$ for all $f \in L^1(G)$ such that $\hat{f}(\gamma) = 0$ for all γ in a neighborhood of Y.

Using the continuity of the functional φ (in (9.1)), we see that Definition 9.1 is equivalent to:

DEFINITION 9.1′: The spectrum of $h \in L^\infty(G)$ is the smallest closed set $Y \subseteq \hat{G}$ such that $f \in J(Y)$ implies $\int fh\,dx = 0$.

For a measure $\nu \in M(\hat{G})$, we see that the support of ν is the same as the spectrum of $\hat{\nu}$.

For a closed set $Y \subseteq \hat{G}$, we let $M(Y)$ be the set of regular Borel measures ν on $\hat{G}$ whose supports are subsets of Y. We let $PM(Y)$ be the set of $h \in L^\infty(G)$ ("pseudomeasures") whose spectra are subsets of Y. Thus $M(Y)\hat{} = \{\hat{\nu} : \nu \in M(Y)\}$ is a subset of $PM(Y)$.

We give $L^\infty(G)$ the weak-* topology, that is, a *net* $\{h_\alpha\} \subseteq L^\infty(G)$ converges, in this topology, to $h \in L^\infty(G)$ if and only if

$$\lim \int fh_\alpha\,dx = \int fh\,dx, \quad \text{all } f \in L^1(G).$$

We may reformulate the notion of spectral synthesis:

THEOREM 9.2: *A closed set $Y \subseteq \hat{G}$ is a set of spectral synthesis if and only if the weak-* closure of $M(Y)$ is all of $PM(Y)$.*

In words, each h with spectrum Y can be "synthesized" by Fourier-Stieltjes transforms of measures concentrated on Y.

We will prove the easy half of 9.2: Suppose $M(Y)$ is weak-* dense in $PM(Y)$. We will show $I(Y) = J(Y)$. Note that if integration against $h \in L^\infty(G)$ annihilates every element of $J(Y)$, then the spectrum of h is a subset of Y. If $f \in I(Y)$ and $\nu \in M(Y)$, then $\hat{f} = 0$ a.e., $d\nu$ so

$$0 = \int \hat{f}\, d\nu.$$

Let $h \in L^\infty(G)$ annihilate $J(Y)$. Then $h \in PM(Y)$, so, by hypothesis, there exists a net $\{\nu_\alpha\} \subseteq M(Y)$ such that $\hat{\nu}_\alpha$ converges to h, weak-$*$ in $L^\infty(G)$. But, then, for $f \in I(Y)$,

$$\int fh\, dx = \lim \int f\hat{\nu}_\alpha\, dx = \lim \int \hat{f} d\nu_\alpha = 0$$

by (9.2), so every $h \in PM(Y)$ annihilates *every* $f \in I(Y)$. If $I(Y) \neq J(Y)$, then there would be an $h \in L^\infty(G)$, (i.e., a φ, by (9.1)), which would annihilate $J(Y)$, but not some $f \in I(Y)$. This would be a contradiction.

10. INTERPOLATION SETS

For a locally compact topological Hausdorff space, X, we denote by $C_0(X)$ the set of continuous functions $f: X \to \mathbf{C}$ such that, for every $\epsilon > 0$, there exists a compact set $C \subseteq X$ such that

$$\sup_{x \in X,\, x \notin C} |f(x)| < \epsilon.$$

DEFINITION 10.1: A closed subset $Y \subseteq \hat{G}$ is an *interpolation set* (or *Helson set*) if for each $f \in C_0(Y)$, there exists $g \in L^1(G)$ such that $\hat{g}(\gamma) = f(\gamma)$, for all $\gamma \in Y$.

Y is a *K-set* (or *Kronecker set*) if for each continuous function $f: Y \to \mathbf{T}$ and $\epsilon > 0$, there exists $x \in G$ such that

$$\sup_{\gamma \in Y} |f(\gamma) - (x, \gamma)| < \epsilon. \tag{10.1}$$

It is not hard to show that every K-set is an interpolation set.

Finite sets are always interpolation sets, since the set of Fourier transforms separates points. In $\mathbf{R}^2$, the set $\{(1, 0), (0, 1)\}$ is a K-set.

In fact, every finite independent set of elements of infinite order $\{\gamma_1, \ldots, \gamma_n\} \subseteq G$ is a K-set. (For γ's in $\mathbf{R}$, this fact is often called "Kronecker's Theorem." A set $\{\gamma_1, \ldots, \gamma_n\} \subseteq G$ is *independent* if $m_1 \in \mathbf{Z}, \ldots, m_n \in \mathbf{Z}$ and $\Sigma_1^n m_j \gamma_j = 0$ imply $m_1\gamma_1 = \cdots = m_n\gamma_n = 0$.) The set $\{10^j : j = 1, 2, \ldots\}$ is an interpolation set in $\mathbf{Z}$; the set $\{0\} \cup \{10^{-j} : j = 1, 2, \ldots\}$ is an interpolation set in $\mathbf{R}$; the set $\{0\} \cup \{j^{-1} : j = 1, 2, \ldots\}$ is *not* an interpolation set. It is a (non-trivial) theorem that interpolation sets cannot contain arbitrarily long arithmetic progressions.

The preceding examples are *not* examples of *perfect* interpolation sets. (A set is "perfect" if it has no isolated points, that is every point of the set is a limit of a net of elements in the set less the point in question.) Perfect interpolation sets do exist. The most striking proof that perfect K-sets exist in $\mathbf{T}$ is that of Robert Kaufman (a definition of the Cantor middle-third set is in Section 12):

THEOREM 10.2: *Let C be Cantor's middle-third set. Let X be the set of continuous maps from C to $\mathbf{T}$ with the metric*

$$d(f, g) = \sup_{x \in C} |f(x) - g(x)|. \tag{10.2}$$

Then, there is a set of first category $X' \subseteq X$, such that if $f \in X, f \notin X'$, then $f(C)$ is a perfect K-set.

(We note that X is a complete metric space so the Baire category theorem says, in particular, that $X' \neq X$.)

Sets of interpolation are often called "Helson sets" because of two theorems Helson proved about them. The first is:

THEOREM 10.3: *Let $Y \subseteq \hat{G}$ be compact. Then the following are equivalent*:

(i) *Y is an interpolation set.*
(ii) *There exists $\alpha > 0$ such that if $f : Y \to \mathbf{C}$ is continuous and has modulus bounded by 1, there exists $g \in L^1(G)$, with $\hat{g} = f$ on Y and $\|g\| \leqslant 1/\alpha$.*

(iii) *There exists* $\alpha > 0$ *such that* $\nu \in M(Y)$ *implies*

$$\sup_{x \in G} |\hat{\nu}(x)| \geqslant \alpha \|\nu\|.$$

The constant α *may be chosen to be the same in* (ii) *and* (iii).

This theorem has an immediate corollary (using the open mapping theorem): *If* $M(Y)\hat{} = PM(Y)$, *then* Y *is an interpolation set.*

Helson's second theorem is:

THEOREM 10.4: *Let* $Y \subseteq G$ *be a compact interpolation set and* $\nu \in M(Y)$, $\nu \neq 0$. *Then* $\hat{\nu} \notin C_0(G)$.

The number α in 10.3 (ii) and (iii) is called the *interpolation* (or the *Helson*) *constant* of Y.

Theorem 10.3 is proved by a simple duality argument.

An immediate application of the Riemann–Lebesgue Lemma and 10.4 show that interpolation sets have zero Haar measure. We gave an example above of a countable set which is not an interpolation set, and so there are sets with zero Haar measure which are not interpolation sets. There are no satisfactory characterizations of interpolation sets in G, except for $G = \mathbf{D}_p\hat{}$ (p prime), in which case the interpolation sets are just the finite unions of independent sets.

11. UNIQUENESS AND SYNTHESIS FOR INTERPOLATION SETS

We begin with an interpretation of Helson's Theorem 10.4. We say a closed set $Y \subseteq \hat{G}$ is a *set of multiplicity in the strong sense* if there exists a non-zero measure $\nu \in M(Y)$ such that $\hat{\nu} \in C_0(G)$.

We say Y is a *set of uniqueness* if whenever $h \in L^\infty(G)$ vanishes at infinity on G and has spectrum contained in Y, then $h = 0$.

If Y is not a set of uniqueness, we say Y is a *set of multiplicity*. Thus, sets of strong multiplicity are sets of multiplicity. (Recall, from Section 9, that the spectrum of ν is the support of the measure ν.)

A discussion of sets of uniqueness appears in the article by Professor Zygmund (pages 1–19) in this volume. Let us point out here that countable sets are sets of uniqueness, and that perfect sets of Haar measure zero may, or may not, be sets of uniqueness, depending on the set.

We see that Helson's Theorem 10.4 says that an interpolation set is not a set of multiplicity in the strong sense.

The following theorem now relates uniqueness and spectral synthesis for interpolation sets:

THEOREM 11.1: *Let $Y \subseteq \hat{G}$ be an interpolation set. The following are equivalent:*

(i) *Y is a set of uniqueness;*
(ii) *Y is a set of spectral synthesis;*
(iii) $M(Y)\hat{} = PM(Y)$.

This is not terribly hard to establish, and shows that two types of "thinness" (interpolation, uniqueness) for subsets of $\hat{G}$ are closely related to the ideal theory of $L^1(G)$, and lends greater interest to the question: Is a set of interpolation necessarily a set of uniqueness? In one case the answer is "yes", and is due to N. Th. Varopoulos:

THEOREM 11.2: *Let Y be a K-set. Then Y is also a set of uniqueness.*

However, this result does not extend to general interpolation sets, as was shown by T. W. Körner in 1970:

THEOREM 11.3: *There exists a compact perfect, independent set $Y \subseteq \mathbf{T}$ such that*

(i) *Y is a set of multiplicity;*
(ii) *Y is an interpolation set with interpolation-constant one.*

(Körner actually gives a stronger version of (ii)).

Körner has constructed a great variety of "thin" sets. One of his constructions shows that the "union theorem" for interpolation sets is sharp:

THEOREM 11.4: *Let Y and $W \subseteq \hat{G}$ be interpolation sets for $L^1(G)$. Then $Y \cup W$ is an interpolation set. If Y and W are both K-sets, then the Helson-constant α for $Y \cup W$ is at most $1/2$.*

Körner's example shows that $\alpha = 1/2$ is sharp in 11.4.

Theorem 11.4 was proved first for discrete groups G by S. W. Drury, and for Y and W K-sets by Varopoulos. Drury's method uses non-classical groups, even when $Y, W \subseteq \hat{G} = \mathbf{Z}$, and Varopoulos' method uses *non* locally compact groups. Varopoulos combined Drury's method with his result for K-sets to prove 11.4 for metrizable $\hat{G}$. Subsequently, Herz gave a simpler proof based on Drury's basic Lemma; although Herz's proof eliminated much of Varopoulos' machinery, non-classical groups are still required, even when $Y, W \subseteq \mathbf{T}$, or $Y, W \subseteq \mathbf{Z}$.

12. CANTOR SETS

This section discusses some relationships between number theory and harmonic analysis.

Let $0 < \xi < 1/2$. We shall construct a set E_ξ as follows: E_ξ shall be the intersection of certain sets $E_0, E_1, \ldots,$

$$E_\xi = \bigcap_{n=0}^{\infty} E_n.$$

We set $E_0 = [0, 1]$, and $E_1 = [0, \xi] \cup [1 - \xi, 1]$. Thus, to make E_1 from E_0, we removed the middle $(1 - 2\xi)$th portion of E_0. We make E_2 from E_1 by removing the middle $(1 - 2\xi)$th from all intervals which compose E_1; so

$$E_2 = [0, \xi^2] \cup [\xi(1 - \xi_1), \xi_2] \cup [1 - \xi, 1 - \xi + \xi^2] \cup [1 - \xi^2, 1].$$

This process is then repeated to produce E_3, then E_4 and so on.

The elements of E_ξ are those numbers $x \in [0, 1]$ which may be written as a sum:

$$x = (1 - \xi) \sum_{j=1}^{\infty} \epsilon_j \xi^j \quad \text{where} \quad \epsilon_j = 0, 1.$$

The choice of the values of ϵ_j varies with x, of course.

Cantor's middle-third set occurs when $\xi = 1/3$.

Before proceeding to results about E_ξ, we make a definition.

DEFINITION 12.1: A number $x \in \mathbf{R}$ is a *Pisot number* if

(i) $x > 1$;

(ii) x is an algebraic integer, that is, there exist $n \geqslant 1$ and integers $a_0, \ldots, a_{n-1}$ such that

$$x^n + \sum_{j=0}^{n-1} a_j x^j = 0 \tag{12.1}$$

and so that the polynomial on the left side of (12.1) is irreducible;

(iii) all other (possibly complex) solutions y of (12.1) have modulus $|y| < 1$.

For example, every integer 2, 3, ... is a Pisot number. A straightforward calculation shows that $(1 + \sqrt{5})/2$ is also a Pisot number. (It is a solution to $x^2 - x - 1 = 0$; the other solution is $(1 - \sqrt{5})/2$ which has absolute value less than 1.)

A surprising fact about Pisot numbers is the following result of Raphaël Salem:

THEOREM 12.2: *The set of Pisot numbers is closed.*

In 1937 Nina Bary proved that if $0 < p/q < 1/2$ is a rational number in reduced form (p and q have no common factor), then $E_{p/q}$ is a set of uniqueness if and only if $p = 1$. In particular, Cantor's middle-third set is a set of uniqueness. In 1943 Salem stated an extension of this result:

THEOREM 12.3: *E_ξ is a set of uniqueness if and only if $1/\xi$ is a Pisot number.*

Note that $1/(1/q) = q$ is a Pisot number, so 12.3 includes Bary's result.

Salem's original proof had an error which he later discovered and partially corrected in 1954. A complete corrected proof of 12.3 was given by Salem and Zygmund in 1955.

In 1956, C. S. Herz proved that Cantor's middle-third set $E_{1/3}$ was a set of spectral synthesis, and in 1970 Yves Meyer proved an extension of this result:

THEOREM 12.4: *If $1/\xi$ is a Pisot number, then E_ξ is a set of spectral synthesis.*

Because a set E_ξ is never an interpolation set, Meyer's result does not follow from 12.3 and 11.1.

It is striking that the proof of 12.4 uses non-classical groups in what appears to be a fundamental way.

(Meyer's proof yields a stronger result than 12.4. Namely, that if $h \in PM(E_\xi)$, then h is a weak-$*$ limit of a sequence of Fourier-Stieltjes transforms of measures ν_n, where ν_n has support contained in the finite set

$$\left\{(1-\xi)\sum_{j=0}^{n} \epsilon_j \xi^j : \epsilon_j = 0, 1, 1 \leqslant j \leqslant n\right\}$$

and

$$\sup_x |\hat{\nu}_n(x)| \leqslant \|h\|, \text{ for all } n.)$$

13. SCHWARTZ'S EXAMPLE

We give the essential idea of Schwartz's proof of Theorem 8.4. We begin with a general discussion of $L^1(\mathbf{R}^3)$. Let Ω denote the set of complex-valued, C^∞-functions f on $\mathbf{R}^3$ which vanish outside a compact set. If $f \in \Omega$, then $\dfrac{\partial f}{\partial x} \in \Omega$ and an integration by parts shows that

$$\left(\frac{\partial f}{\partial x}\right)^{\wedge}(x, y, z) = -ix\hat{f}(x, y, z). \tag{13.1}$$

Repeating this, we see that

$$\left(\frac{\partial^2 f}{\partial x^2}\right)^{\wedge}(x, y, z) = -x^2\hat{f}(x, y, z),$$

and by changing the variable of differentiation and computing, we find:

$$\sup_{x, y, z} |\hat{f}(x, y, z)|(x^2 + y^2 + z^2)^2 < \infty.$$

This proves that if $f \in \Omega$, then $\hat{f} \in L^1(\mathbf{R}^3)$. It is not hard to show that $\{\hat{f} : f \in \Omega\}$ is dense in $L^1(\mathbf{R}^3)$ (but see below). We will construct a linear functional L on $L^1(\mathbf{R}^3)$ such that $L(\hat{f}) = 0$ if $f \in \Omega$, and $\hat{f}$ is zero in a neighborhood of the unit sphere $S = \{(x, y, z) : x^2 + y^2 + z^2 = 1\}$.

Let ν denote surface area measure on S. If $f \in \Omega$, we define $L(\hat{f})$ by:

$$L(\hat{f}) = \int \frac{\partial f}{\partial x}\, d\nu. \tag{13.2}$$

(L is called a "distribution derivative" of ν.) We now show that $|L(\hat{f})| \leqslant 4\pi\|\hat{f}\|$, so that L will have a unique extension to a continuous linear functional on $L^1(\mathbf{R}^3)$. (Here, $\|\hat{f}\|$ is the norm of $\hat{f}$ in $L^1(\mathbf{R}^3)$.)

An application of Fubini's theorem to (13.2) shows

$$\begin{aligned} L(\hat{f}) &= \int \hat{\nu}(x, y, z)\left(\frac{\partial f}{\partial x}\right)^{\wedge}(x, y, z)\, dx\, dy\, dz \\ &= -\int \hat{\nu}(x, y, z) ix\hat{f}(x, y, z)\, dx\, dy\, dz. \end{aligned}$$

Therefore

$$|L(\hat{f})| \leqslant \sup_{x, y, z} |x\hat{\nu}(x, y, z)| \, \|\hat{f}\|.$$

We now compute $\hat{\nu}$:

$$\hat{\nu}(x, y, z) = \int e^{-i(xx' + yy' + zz')} \, d\nu(x', y', z'). \tag{13.3}$$

Because ν is a spherically symmetric distribution of mass, $\hat{\nu}(x, y\ z)$ depends only on $x^2 + y^2 + z^2 = r^2 \geqslant 0$, the square of the distance of (x, y, z) from $(0, 0, 0)$. We may therefore assume $y = z = 0$ and $x = r$. We put (13.3) in spherical coordinates, using x as the polar axis and θ as the polar angle, and integrate, using the definition of ν:

$$\hat{\nu}(r, 0, 0) = \int_0^{2\pi} \int_0^{\pi} e^{ir \cos \theta} \sin \theta \, d\theta \, d\phi$$

$$= \frac{4\pi \sin r}{r}.$$

Thus

$$|x\hat{\nu}(x, y, z)| \leqslant 4\pi \sin r$$

where $r = (x^2 + y^2 + z^2)^{1/2}$, and this proves that

$$|L(\hat{f})| \leqslant 4\pi \|\hat{f}\|,$$

so L does extend to a bounded linear functional on $L^1(\mathbf{R}^3)$.

We let J be the closure in $L^1(\mathbf{R}^3)$ of $\{\hat{f} : f \in \Omega, f = 0$ in a neighborhood of $S\}$ and I be the closure of $\{\hat{f} : f \in \Omega, f = 0$ on $S\}$. It is easy to see that $I \supseteq J$, that I and J are ideals, and that $L(g) = 0$ for all $g \in J$. The reader may construct a $g \in I$ such that $L(g) \neq 0$ (an appropriate $f \in \Omega$, with $\hat{f} = g$, will do). The proof is now almost complete. All that is needed are the observations that $L(f)$ depends only on the behavior of $\hat{f}$ in a neighborhood of S and that $J = J(S)$. The second follows from the first, which is fairly easily established.

PART THREE. BRIEF SUGGESTIONS FOR FURTHER READING.

Sections 1–5 follow very closely the (much more detailed) presentation of Chapters 1–3 of the book of Walter Rudin [15], which is a good introduction to the harmonic analysis on abelian groups for readers who have some familiarity with results for the line or circle. The book of Yitzak Katznelson [8] is a good book for readers who are not familiar with $L^1(\mathbf{R})$; it combines the abstract viewpoint with hard calculations, and provides instructive exercises of varying difficulty. Each of these books contains a proof of the Wiener-Pitt Theorem (our Section 7). For a modern approach in the spirit of Weil's [16], see the encyclopedic work of Hewitt and Ross [6].

For Banach algebras, the original paper [3] of Gelfand is still informative, as is [4]. Each of the books [2, 12, 14] also provides a good introduction to Banach algebra theory.

The reader who is interested in the group aspect of harmonic analysis might read the classical and still interesting books of Pontryagin [13] and Weil [16], or Volume I of [6]. The book of Nachbin [11] is a very leisurely and readable discussion (with several proofs) of the existence of Haar measure.

The book of Yves Meyer [10] is probably the best single source for the material in part two. The books of Bary [1] and Zygmund [18] contain good sections on sets of uniqueness, and are well worth reading in any case. For constructions of sets E_ξ, [7] is comprehensive and clear and contains many of the results in Part Two. A continuation of [7] is the Seminar report [9] which contains proofs of Kaufman's theorem (10.2 above) and the union theorem (11.4). (A better proof of Theorem 11.4 appears in the article [5] of Carl Herz.)

REFERENCES

1. Bary, N., *Trigonometric Series*, Macmillan, London, 1964.
2. Bonsall, F. F., and J. Duncan, *Complete Normed Algebras*, Springer-Verlag, New York, 1973.

3. Gelfand, I. M., "Normierte ringe," *Mat. Sb.*, **9** (1941), 3–24.

4. ——, , D. A. Raikov, and G. E. Silov, "Commutative normed rings," *Usp. Mat. Nauk*, **1** (1946), **2** (12), 48–146; *Amer. Math. Soc. Transl.* (2) **5** (1957), 115–220.

5. Herz, C. S., "Drury's lemma and Helson sets," *Studia Math.*, **42** (1972), 205–219.

6. Hewitt, E., and K. A. Ross, *Abstract Harmonic Analysis*, I, II, Springer-Verlag, New York, 1963, 1970.

7. Kahane, J.-P., and R. Salem, *Ensembles Parfaits et Séries Trigonométriques*, Hermann, Paris, 1963.

8. Katznelson, Y., *An Introduction to Harmonic Analysis*, Wiley, New York, 1968.

9. Lindahl, L. A., and F. Poulsen, *Thin Sets in Harmonic Analysis*, Marcel Dekker, New York, 1971.

10. Meyer, Y., *Algebraic Numbers and Harmonic Analysis*, North-Holland, Amsterdam, 1972.

11. Nachbin, L., *The Haar Integral*, Van Nostrand, Princeton, N. J., 1965.

12. Naimark, M. A., *Normed Rings*, Noordhoff, Groningen, 1959–64; *Normed Algebras*, (2nd Ed. of *Normed Rings*), Noordhoff, 1972.

13. Pontryagin, L. S., *Topological Groups*, Gordon and Breach, New York, 1966.

14. Rickart, C. E., *General Theory of Banach Algebras*, Van Nostrand, Princeton, 1960.

15. Rudin, W., *Fourier Analysis on Groups*, Wiley, New York, 1962.

16. Weil, A., *L'intégration dans les Groupes Topologiques et ses Applications*, Hermann, Paris, 1938 (reprinted 1951).

17. Weiss, G., "Harmonic Analysis," *Studies in Real and Complex Analysis, M.A.A.Studies in Mathematics*, Vol. 3, Mathematical Association of America, Washington, DC, 1965.

18. Zygmund, A., *Trigonometric Series, vol. I, II*, Cambridge University Press, Cambridge, England, 1968.

HARMONIC ANALYSIS ON COMPACT GROUPS

Guido Weiss

1. THE PETER-WEYL THEOREM

Our purpose will be to describe some of the basic features of harmonic analysis on compact groups. In order to do so we will have to know something about their structure. This first section is devoted to a presentation of the most relevant aspects of this structure. The second section will be mostly involved with specific examples of results that extend classical theorems in the theory of Fourier series. In general, we shall not include proofs. The third and last section will include a short guide to the literature and, in particular, will lead the reader to places where such proofs may be found and, also, will give him or her some idea of the directions harmonic analysis on compact groups has taken.

A *compact topological group* G is a group endowed with a compact Hausdorff topology in such a way that the mapping $(u, v) \to uv^{-1}$, from $G \times G$ into G, is continuous. The finite groups, the group, $U(n)$, of $n \times n$ unitary matrices and the group, $O(n)$, of $n \times n$ orthogonal matrices furnish us with familiar examples of compact groups. The classical theory of Fourier series involves the *circle group* (or the *one-dimensional torus*), T, consisting of the real numbers modulo 1. In order to motivate our study we first examine some aspects of this theory from a point of view that is somewhat different from the one commonly used but does serve as a natural introduction to the extension of those notions that will be presented now.

The group operation of T induces the family of *translation operators* $\{R_\theta : \theta \in T\}$, acting on function on T, defined by

$$(R_\theta f)(\varphi) = f(\varphi - \theta).$$

One of the basic features of classical harmonic analysis is that almost all the operators that arise commute with translations. For example, differentiation has this commutativity property. This is also clearly true of the convolution operators; that is, those operators mapping an integrable f into the function whose value at φ is

$$(f*g)(\varphi) = \int_0^1 f(\varphi - \theta)g(\theta)\,d\theta,$$

where g is some (fixed) integrable function on T. For example, when

$$g(\theta) = D_n(\theta) = \frac{\sin(2n+1)\pi\theta}{\sin \pi\theta}$$

is the *Dirichlet kernel* we obtain the mapping of f into the nth partial sum of its Fourier series. Similarly, the Abel and Cesáro means of the Fourier series of f are obtained from convolution operators.

In classical harmonic analysis a key role is played by the exponential functions

$$e_k(\theta) = e^{2k\pi i\theta}, \tag{1.1}$$

$k = 0, \pm 1, \pm 2, \ldots$. Their importance arises from the fact that they provide a common spectral decomposition for the operators that commute with translations. Let us make this last assertion more precise. Suppose A is a linear operator defined on a space of functions on T that includes the *trigonometric polynomials* (i.e., the finite linear combinations of the exponential functions (1.1)). We assume

$$\text{(i)} \quad AR_\theta = R_\theta A \quad \text{for all } \theta \in T; \text{ (ii)} \quad Ae_k \textit{ is continuous} \tag{1.2}$$

for all $k = 0, \pm 1, \pm 2, \ldots$. Since the exponential functions satisfy

$$e_k(\theta + \varphi) = e_k(\theta)e_k(\varphi) \tag{1.3}$$

for all $\theta, \varphi \in T$, it follows that each e_k is an eigenfunction of all the operators R_θ. Indeed,

$$R_{-\theta}e_k = e_k(\theta)e_k.$$

Thus, if A is a linear operator on the trigonometric polynomials, then $A(R_{-\theta}e_k) = e_k(\theta)Ae_k$. If, moreover, A satisfies (1.2), then

$$(Ae_k)(\theta) = (R_{-\theta}Ae_k)(0) = (AR_{-\theta}e_k)(0) = e_k(\theta)(Ae_k)(0).$$

That is,

$$Ae_k = a_k e_k \tag{1.4}$$

with $a_k = (Ae_k)(0)$. In other words, each operator A satisfying (1.2) has the exponential functions (1.1) as a system of eigenfunctions (proper vectors). Since this system is complete (in various senses; for example, in $L^2(0, 1)$), $\{e_k\}$ is a complete system of proper functions of A. The corresponding sequence of eigenvalues $a_k = (Ae_k)(0)$, $k = 0, \pm 1, \pm 2, \ldots,$ will be called the sequence of *Fourier coefficients* of A. We should observe that if A is a convolution operator induced by some fixed integrable function g, then

$$\begin{aligned} a_k &= (Ae_k)(0) \\ &= \int_0^1 g(\theta)e_k(0 - \theta)\, d\theta \\ &= \int_0^1 g(\theta)e^{-2\pi i k\theta}\, d\theta = \hat{g}(k) \end{aligned}$$

is the classical kth Fourier coefficient of the function g. If, say, A is the differential operator $d/d\theta$, then $a_k = 2\pi ik$.

When A satisfies (1.2), its Fourier transform is, by definition, the sequence $\{(\mathfrak{F}A)(k)\}$ of the restrictions of A to the (one-dimensional) spaces generated by e_k, $k = 0, \pm 1, \pm 2, \ldots$. The Fourier coefficients of A are then the (one-by-one) matrices of $(\mathfrak{F}A)(k)$ with respect to the basis e_k, $k = 0, \pm 1, \pm 2, \ldots$. This definition gives us the following interpretation of the well-known formula

$$(f*g)\,\hat{}\,(k) = \hat{f}(k)\hat{g}(k), \tag{1.5}$$

valid for integrable functions f and g: the operator induced by convolution with $f*g$ is the composition of the convolution operators induced by g and by f. Moreover, an operator A satisfying (1.2) is completely determined by the sequence of its Fourier coefficients in case the trigonometric polynomials are dense in an appropriate sense in the domain of A; also, given such a sequence $\{a_k\}$, equality (1.4) and linearity can be used to define A on the space of trigonometric polynomials. In this case we write

$$A \sim \sum_{k=-\infty}^{\infty} a_k e_k \tag{1.6}$$

and refer to the series on the right as the *Fourier series expansion* on A. We shall also call such an operator *a generalized convolution operator*.

The main purpose of this section is to show how these properties can be extended to the general compact group G. It turns out that one can describe these extensions in a simple manner. There are, however, some difficulties that are encountered immediately. For example, the fact that we do not assume G to be commutative leads us (in general) to two different classes of translation operators: the left translations $\{L_u : u \in G\}$ defined by

$$(L_u f)(y) = f(u^{-1}v),$$

and the right translations $\{R_u : u \in G\}$ defined by

$$(R_u f)(v) = f(vu),$$

whenever f is a function whose domain is G and $v \in G$. Let us observe, in passing, that by employing the inverse of u in the definition of L_u we obtain the homomorphism properties

$$L_{u_1u_2} = L_{u_1}L_{u_2}, \qquad R_{u_1u_2} = R_{u_1}R_{u_2}, \tag{1.7}$$

for all $u_1, u_2 \in G$. It is natural to suspect that there is a difference between the operators that commute with left translations and the ones that commute with right translations and that those operators commuting with both left and right translations should play a particularly important role in the study of harmonic analysis on G. This is indeed the case and it will become increasingly more evident as we develop the subject.

Lebesgue measure on T has the well-known analog of Haar measure on G (so normalized that the measure of G is 1). Since the group is compact, Haar measure is invariant under both left and right translations. We shall denote the element of Haar measure by du and write

$$\int_G f(u)\,du = \int f$$

for the Haar integral of $f \in L^1(G)$. Now, given a function $g \in L^1(G)$, it induces two convolution operators on $L^1(G)$: the *left convolution operator* $A^l = A_g^l$ defined by

$$\begin{aligned}(A^l f)(u) &= (g*f)(u)\\ &= \int_G g(v^{-1}u)f(v)\,dv\\ &= \int_G g(v)f(uv^{-1})\,dv\end{aligned}$$

and the *right convolution operator* $A^r = A_g^r$ defined by

$$\begin{aligned}(A^r f)(u) &= (f*g)(u)\\ &= \int_G g(uv^{-1})f(v)\,dv\\ &= \int_G g(v)f(v^{-1}u)\,dv.\end{aligned}$$

From the position of the variable u in the last integrals defining the two convolution operators we see immediately that

$$L_w A^l = A^l L_w \quad \text{and} \quad R_w A^r = A^r R_w \tag{1.8}$$

for all $w \in G$. It follows, therefore, that if we are seeking extensions of the notion of a generalized convolution operator, it is natural to consider separately those operators commuting with left translations and those commuting with right translations.

In order to explore the extension of property (1.2), we need an analog of the family of exponential functions (1.1). As we shall see later, a rather complete analog exists when G is commutative. Unfortunately, when this is not the case there is no common spectral decomposition for operators satisfying either of the commutativity relations (1.8). That is, one cannot find an orthogonal decomposition of $L^2(G)$ in terms of *one-dimensional* subspaces of "nice" functions that are mapped into themselves by such operators. The general situation, however, is not very different from the one we described for T. It turns out that $L^2(G)$ decomposes into an orthogonal direct sum of finite dimensional spaces $\mathcal{H}_\lambda$, $\lambda \in \Lambda$, having the following properties (the nature of the indexing set Λ will be discussed later):

(i) $\mathcal{H}_\lambda$ *consists of continuous functions;*

(ii) *if a linear operator A maps each $\mathcal{H}_\lambda$ into continuous functions on G and commutes with either $\{L_u : u \in G\}$ or $\{R_u : u \in G\}$, then A maps $\mathcal{H}_\lambda$ into itself;**

(iii) *no proper subspace of $\mathcal{H}_\lambda$ is mapped into itself by the collection of left and right translations by elements of G.*

Thus, the spaces $\mathcal{H}_\lambda$ correspond to the one-dimensional spaces generated by the exponential functions e_k. Property (ii) is the generalization of equality (1.4). The collection $\{\mathcal{H}_\lambda : \lambda \in \Lambda\}$ is the set of *all* minimal left and right invariant subspaces of $L^2(G)$. Thus, the direct sum decomposition

$$L^2(G) = \sum_{\lambda \in \Lambda} \oplus \mathcal{H}_\lambda \tag{1.9}$$

---

*The obvious commutativity relation $R_u L_v = L_v R_u$ implies, by (ii), that L_v and R_u map $\mathcal{H}_\lambda$ into itself. (iii) asserts that $\mathcal{H}_\lambda$ is minimal with respect to this property.

is unique. If G is abelian, each $\mathcal{H}_\lambda$ is one-dimensional (as in the case when $G = T$).

It is remarkable that so much is implied by the few and simple assumptions that consist in the definition of a compact topological group. Actually, one can obtain a considerably more detailed description of the structure of such groups. We shall now endeavor to present the most important features of this description. These properties are usually referred to as the *Peter–Weyl theorem*.

In order to give this account and to see how the spaces $\mathcal{H}_\lambda$ arise, we need to consider the (unitary) *representations* of G. By such a representation we mean a continuous mapping S from G into the collection of unitary operators acting on a Hilbert space H satisfying the homomorphism property

$$S_{u_1u_2} = S_{u_1}S_{u_2} \tag{1.10}$$

for all $u_1, u_2 \in G$. Equalities (1.7) and the properties of Haar measure show that the operators L_u, R_u, $u \in G$, give rise to unitary representations $L : u \to L_u$ and $R : u \to R_u$ acting on the Hilbert space $H = L^2(G)$. These two mappings are called the *left regular* and the *right regular* representations of G.

If S and S' are two representations of G acting on the Hilbert spaces H and H', then S is said to be equivalent to S' if and only if there exists a unitary transformation $U : H' \to H$ such that $U^{-1}S_vU = S'_v$ for all $v \in G$. The representation S is said to be *irreducible* if there does not exist a proper subspace $K \subset H$ such that S_u maps K into itself for all $u \in G$. If such a subspace K exists, it is called an *invariant subspace* and the representation S is said to be *reducible*; since S_u is unitary it follows easily that the orthogonal complement, $K^\perp$, of K is also an invariant subspace. It is now natural to inquire whether the restrictions of S_u, $u \in G$, to these subspaces give rise to reducible or irreducible representations (called the restriction of S to K and $K^\perp$). If both representations are irreducible, we have the orthogonal direct sum decomposition $H = K \oplus K^\perp$ on which the restrictions of S act irreducibly. If this is not the case, we can decompose H further. We are thus led to ask whether there is an orthogonal decomposition of H whose summands are minimal invariant subspaces under the action of S.

The answer to this question is that such a decomposition always does exist, but it is, in general, an infinite one. In particular, there exist orthogonal direct sum decompositions of $L^2(G)$ such that the restrictions of *either* the left or the right regular representation to the various summands consist of *all* the irreducible representations of G (up to equivalency).

It is not hard to see why the last assertion is true. Suppose S is an irreducible representation of G acting on the Hilbert space H. It can be shown that H has a finite dimension d. We can, therefore, select an orthonormal basis of H and consider the matrix

$$e(u) = \begin{bmatrix} t_{11}(u) & t_{12}(u) & \cdots & t_{1d}(u) \\ t_{21}(u) & t_{22}(u) & \cdots & t_{2d}(u) \\ \cdots & \cdots & \cdots & \cdots \\ t_{d1}(u) & t_{d2}(u) & \cdots & t_{dd}(u) \end{bmatrix}$$

of S_u with respect to this basis. Making use of an elementary but very important result in group theory, called Schur's lemma, one can prove that the functions t_{ij}, $1 \leqslant i, j \leqslant d$, (the entries of the matrix e we just introduced) satisfy

$$\int_G t_{ij}(u)\,\overline{t_{kl}(u)}\;du = \frac{\delta_{ik}\delta_{jl}}{d}, \tag{1.11}$$

where δ_{ik} is the Kronecker δ function (equal to 1 if $i = k$ and equal to 0 otherwise).* That is, these functions are *mutually orthogonal and each has norm* $1/\sqrt{d}$. They span, therefore, a d^2-dimensional subspace of $L^2(G)$. In fact, *this is one of the spaces* $\mathcal{H}_\lambda$ *occurring in the direct sum* (1.9).

Since S is a representation, we know that each matrix $e(u)$ is unitary and, moreover,

$$e(uv) = e(u)e(v). \tag{1.12}$$

---

*Since the representation S is assumed to be continuous, it follows that each t_{ij} is a continuous function. Thus, the integrals (1.11) are well defined.

This equality is obviously an extension of the functional equation (1.3) satisfied by the exponential functions on T. In terms of the coefficients t_{ij} we can write (1.12) in the form

$$R_v t_{ij} = \sum_{k=1}^{d} t_{kj}(v) t_{ik}. \tag{1.13}$$

We see immediately from this relation that the right regular representation maps the ith row space of e into itself. Moreover, (1.13) tells us that the matrix of the restriction of R_v to this row space, with respect to the orthogonal basis $\{t_{i1}, t_{i2}, \ldots, t_{id}\}$ is precisely $e(v)$! Since S was assumed to be irreducible, it follows that this restriction of the right-regular representation acts irreducibly and is a representation equivalent to S.

Similar results hold for the left-regular representation. Because of the use of the inverse, u^{-1}, in the definition of L_u, the column spaces of e play the role (*vis-à-vis* L) just described for its row spaces (*vis-à-vis* R).

If G is abelian, the matrices $e(u)$, $u \in G$, form a commuting family of unitary operators and, thus, they have a common complete set of eigenvectors. The space generated by each one of these vectors is, by definition, mapped into itself by all the $e(u)$'s. If there were two or more independent eigenvectors, then $e(u)$, $u \in G$, could not arise from an irreducible representation. We see, therefore, that the dimension d is 1 when G is abelian as is the case for the special case $G = T$.

When $G = T$, the exponential functions e_k generate the irreducible spaces on which the representations of G act. We obtain all the irreducible representations (up to equivalence) by restricting translations to the span of e_k, $k = 0, \pm 1, \pm 2, \ldots$. It is clear that if $k \neq j$, we obtain inequivalent representations, since R_θ restricted to the span of e_k has the eigenvalue $e^{-ik\theta}$ which cannot equal the eigenvalue, $e^{-ij\theta}$, of the restriction of R_θ to e_j. The situation is similar in the general case. We have shown how to obtain one of the spaces $\mathcal{H}_\lambda$ and have described how the right (left) regular representation acts on it. The representation R restricted to the orthogonal complement (in $L^2(G)$) of this space $\mathcal{H}_\lambda$ can be broken down into irreducible components, but none is equivalent to the restriction of R to one of the row spaces in $\mathcal{H}_\lambda$ we described above. We see, therefore, how a transfinite argument can be employed to obtain the decomposition (1.9) and that the indexing set Λ can be chosen to be the collection of equivalence classes of the irreducible representation of G.

We can now extend the various notions and relations we discussed in connection with T. A *generalized trigonometric polynomial* on a compact group G is a finite sum of functions belonging to the various spaces $\mathcal{H}_\lambda$. A *generalized right* (*left*) *convolution operator* A is a linear operator defined on the generalized trigonometric polynomials with values that are continuous functions on G and satisfying $R_u A = AR_u (L_u A = AL_u)$ for all $u \in G$. The *Fourier transform* of A is the family $\{(\mathfrak{F}A)(\lambda)\}$ of restrictions of A to the spaces $\mathcal{H}_\lambda$, $\lambda \in \Lambda$.

We have just indicated that, associated with each λ, we have a matrix-valued function $e_\lambda = (t_{ij}^\lambda)$ satisfying the functional relationship (1.12) whose entries form a basis of the d_λ^2-dimensional space $\mathcal{H}_\lambda$ for which the orthogonality relations (1.11) hold (with d replaced by d_λ). If A is a generalized right convolution operator the matrix $\hat{A}(\lambda)$, with respect to this basis, of the operator $(\mathfrak{F}A)(\lambda)$ will be called the (*matricial*) *Fourier coefficient* of A. Since the basis $\{t_{ij}^\lambda\}$ is not unique, neither are these Fourier coefficients; however, the operator $(\mathfrak{F}A)(\lambda)$ depends only on λ, G and A. It is not hard to compute the Fourier coefficients of the right convolution operator A_g^r when $g \in L^1(G)$. In fact, let us introduce, in complete analogy with the classical case, the matrix

$$\hat{g}(\lambda) = \int_G g(v) e_\lambda(v^{-1})\, dv.$$

If we order our basis according to the columns of e_λ(t_{ij}^λ precedes t_{kl}^λ if $j < l$ or, when $j = l$, $i < k$), the matrix $\hat{A}_g^r(\lambda)$ has d_λ blocks $\hat{g}(\lambda)$ along the diagonal and zeros elsewhere:

$$\hat{A}_g^r(\lambda) = \begin{bmatrix} \hat{g}(\lambda) & 0 & \cdots & 0 \\ 0 & \hat{g}(\lambda) & \cdots & 0 \\ \cdots & \cdots & \cdots & \cdots \\ 0 & 0 & \cdots & \hat{g}(\lambda) \end{bmatrix}. \qquad (1.14)$$

If $g \in L^2(G)$, we have the Fourier expansion (in the L^2 sense)

$$g = \sum_{\lambda \in \Lambda} d_\lambda \quad \text{trace} \quad (\hat{g}(\lambda) e_\lambda) \tag{1.15}$$

and the Plancherel formula

$$\|g\|_2 = \left(\sum_{\lambda \in \Lambda} d_\lambda ||| \hat{g}(\lambda) |||^2 \right)^{1/2}, \tag{1.16}$$

where $||| \hat{g}(\lambda) |||$ denotes the Hilbert-Schmidt norm (the square root of the sum of the squares of the coefficients of the matrix $\hat{g}(\lambda)$). Suppose f and g are integrable functions on G; then a simple calculation shows that

$$(f * g) = \sum_{\lambda \in \Lambda} d_\lambda \quad \text{trace} \quad (\hat{f}(\lambda) \hat{g}(\lambda) e_\lambda).^* \tag{1.17}$$

This relation is obviously an extension of equality (1.5). Moreover, it describes completely how the operators A_f^l and A_g^r affect the Fourier expansion of the functions g and f. It is not hard to show that the generalized convolution operators act (formally) in exactly the same way on the functions in their domain. More specifically, suppose A is a generalized right convolution operator then, for each $\lambda \in \Lambda$, there exists a $d_\lambda \times d_\lambda$ matrix $m(\lambda)$ such that

$$\hat{A}(\lambda) = \begin{bmatrix} m(\lambda) & 0 & \cdots & 0 \\ 0 & m(\lambda) & \cdots & 0 \\ \cdots & \cdots & \cdots & \cdots \\ 0 & 0 & \cdots & m(\lambda) \end{bmatrix}; \tag{1.18}$$

moreover, if g is a trigonometric polynomial (that is, the Fourier expansion (1.15) is finite), then

$$Ag = \sum_{\lambda \in \Lambda} d_\lambda \quad \text{trace} \quad (\hat{g}(\lambda) m(\lambda) e_\lambda).\dagger \tag{1.19}$$

---

*Since the convolution of two integrable functions is a continuous function, we can interpret the equality (1.17) in the L^2 sense.

†Since the Fourier expansion of g is finite, the sum in (1.19) must also be finite; thus, the equality can be interpreted in the pointwise sense.

In particular, we see that Ag is a trigonometric polynomial whenever g is a trigonometric polynomial. We shall call the collection $\{m(\lambda), \lambda \in \Lambda\}$ the family of *right multiplier matrices associated* with the operator A. (These matrices, of course, depend on the choice of the basis $\{t_{ij}^{\lambda}\}$ spanning $\mathcal{H}_\lambda$). Similarly, we can define the family of *left multiplier matrices associated* with a generalized left convolution operator; the basic difference between these two families is that the formula corresponding to equality (1.19), for a left convolution operator, has the order of multiplication of the matrices $\hat{g}(\lambda)$ and $m(\lambda)$ reversed.

The summands in the series (1.15) are, of course, the projections $P_\lambda g$ of $g \in L^2(G)$ onto the subspaces $\mathcal{H}_\lambda$. Let us examine these projections in more detail. We have

$$(P_\lambda g)(u) = d_\lambda \quad \text{trace} \quad \left\{ \int_G g(v) e_\lambda(v^{-1}) e_\lambda(u)\, dv \right\}$$

$$= d_\lambda \int_G g(v) \quad \text{trace} \quad \left\{ e_\lambda(v^{-1}u) \right\} dv.$$

If we denote trace (e_λ) by χ_λ, we have, therefore,

$$(P_\lambda g)(u) = d_\lambda (g * \chi_\lambda)(u). \tag{1.20}$$

The function χ_λ is called the *character* associated with the equivalence class of irreducible representations λ. Suppose $e_\lambda^{(i)}$, $i = 1, 2$, is a matrix whose entries form an orthogonal basis for $\mathcal{H}_\lambda$, and is obtained from an irreducible representation S^i in the manner described above. Then $e_\lambda^{(1)} = p e_\lambda^{(2)} p^{-1}$, where p is a $d_\lambda \times d_\lambda$ unitary matrix, and it follows that the traces of $e_\lambda^{(1)}$ and $e_\lambda^{(2)}$ are equal. Thus, the characters χ_λ depend only on the class λ. Moreover, using elementary properties of the trace and relation (1.12) we have

$$f * \chi_\lambda = \chi_\lambda * f \tag{1.21}$$

whenever $f \in L^1(G)$, and

$$\chi_\lambda(vuv^{-1}) = \chi_\lambda(u) \tag{1.22}$$

for all $u, v \in G$. In particular, it follows that the Fourier expansion (1.15) can be rewritten in the form

$$g = \sum_{\lambda \in \Lambda} d_\lambda (g * \chi_\lambda)$$

$$= \sum_{\lambda \in \Lambda} d_\lambda (\chi_\lambda * g). \tag{1.23}$$

A function f on G satisfying $f(vuv^{-1}) = f(u)$ for all $u, v \in G$ is called a *central* function. If f is a central integrable function, it follows immediately that $f*g = g*f$ for all $g \in L^1(G)$. From this commutativity relation and equality (1.17) we then have $\hat{f}(\lambda)\hat{g}(\lambda) = \hat{g}(\lambda)\hat{f}(\lambda)$ for all $\lambda \in \Lambda$. But g can be chosen to be an arbitrary linear combination of the basis $\{t_{ij}^\lambda\}$ and, thus, $\hat{g}(\lambda)$ is an arbitrary $d_\lambda \times d_\lambda$ matrix. It follows that $\hat{f}(\lambda)$ must be a scalar matrix. Using this fact, and the orthogonality relations (1.11) and (1.15), we see that *if $f \in L^2(G)$ is a central function, its Fourier expansion is*

$$\sum_{\lambda \in \Lambda} a_\lambda \chi_\lambda, \tag{1.24}$$

where $a_\lambda = \int_G f(v)\chi_\lambda(v^{-1})\,dv$. Because of (1.22), it is also clear that a Fourier expansion of the type in (1.24) must represent a central function. More generally, if A is a generalized (left or right) convolution operator with scalar multiplier matrices $a_\lambda I_\lambda$ (I_λ denotes the $d_\lambda \times d_\lambda$ identity matrix) we shall call it a *central convolution operator*. The action of A on a trigonometric polynomial g is given by the following simplified version of formula (1.19):

$$Ag = \sum_{\lambda \in \Lambda} a_\lambda d_\lambda \ \text{trace} \ (\hat{g}(\lambda)e_\lambda). \tag{1.25}$$

That is, $(AP_\lambda)g = a_\lambda P_\lambda g$ and, thus, each function in $\mathcal{H}_\lambda$ is an eigenvector of A.

We have now presented enough details about the structure of compact groups in order to describe some results in harmonic analysis involving them.

2. SOME EXAMPLES OF HARMONIC ANALYSIS ON COMPACT GROUPS

An important feature of compact groups is their connection with various classes of special functions. The trigonometric functions, the polynomials of Legendre, Jacobi and Gegenbauer and spherical harmonics are examples of such classes arising from the study of classical compact groups. The reason for this is that the generalized trigonometric polynomials we introduced in the first section can often be represented in terms of such functions. A considerable amount of analysis on compact groups has been developed involving this interplay between representation theory and the classical theories of special functions. It would be virtually impossible to present here anything resembling a fair account of this analysis. Our aim is much more modest: we shall try to describe the *harmonic* analysis on compact groups.

At this point, it would be appropriate to say something about what we mean by the term "harmonic analysis". Rather than try to give a definition of this term that would encompass all the aspects of this subject, we claim that, in the context of compact groups, a very large part of what is called "harmonic analysis" is the study of the properties of generalized convolution operators.

This certainly is true of the classical theory of Fourier series where the underlying group is T. Let us examine topics considered in Zygmund's "Trigonometric series": Convergence problems can be phrased in terms of the sequence of convolution operators induced by the sequence of Dirichlet kernels. Similarly, Cesáro (or Abel) summability theory study convolution with the Fejér (or Poisson) kernels. The generalized convolution operator with kth Fourier coefficient $-i \operatorname{sgn} k$ maps an integrable function into its conjugate function. Fractional integration and differentiation are obtained from the convolution operators having fixed powers of $|k|$ for their kth Fourier coefficient.

In the examples just cited, one is usually interested either in the convergence properties of sequences of generalized convolution operators or in studying mapping properties of such an operator (for example, is it bounded as a transformation from $L^p(T)$ to $L^q(T)$ for appropriate values of p and q?). Let us now consider some specific examples of generalized convolution operators acting on functions defined on the group $SU(2)$. This group, called the 2×2 *special unitary group*, consists of the unitary matrices

$$u = \begin{pmatrix} u_{11} & u_{12} \\ u_{21} & u_{22} \end{pmatrix}$$

with determinant $u_{11}u_{22} - u_{12}u_{21} = 1$. The circle group T can be considered to be the group of complex numbers of absolute value 1. Since the complex number field $\mathbf{C}$ is often identified with the two-dimensional Euclidean plane $\mathbf{R}^2$, T can also be regarded, geometrically, as the perimeter of the circle of radius one about the origin in $\mathbf{R}^2$. $SU(2)$ plays the same roles *vis-à-vis* complex two-dimensional space $\mathbf{C}^2$ and four dimensional Euclidean space $\mathbf{R}^4$. To see this, we write the general element $u \in SU(2)$ in the form

$$\begin{aligned} u &= -i\begin{pmatrix} -x_3 + ix_4 & x_1 + ix_2 \\ x_1 - ix_2 & x_3 + ix_4 \end{pmatrix} \\ &= -i\begin{pmatrix} -\overline{z_2} & z_1 \\ \overline{z_1} & z_2 \end{pmatrix}, \end{aligned} \tag{2.1}$$

where $x = (x_1, x_2, x_3, x_4)$ satisfies $|x| \equiv (x_1^2 + x_2^2 + x_3^2 + x_4^2)^{1/2} = 1$ and $z_1 = x_1 + ix_2$, $z_2 = x_3 + ix_4$ (that this can be done follows easily from the hypotheses that u is a unitary matrix of determinant 1). We thus obtain a correspondence $u \leftrightarrow x$, between the elements of $SU(2)$ and points of $\mathbf{R}^4$, as well as a correspondence $u \leftrightarrow z = (z_1, z_2)$ between $SU(2)$ and a subset of $\mathbf{C}^2$. In the first case $SU(2)$ is mapped onto the surface of the unit sphere $\Sigma_3 = \{x \in \mathbf{R}^4 : |x| = 1\}$ and in the second case $SU(2)$ is mapped onto the "complex unit sphere" $\{z = (z_1, z_2) \in \mathbf{C}^2 : z_1\overline{z_1} + z_2\overline{z_2} = 1\}$. Moreover, by writing the matrix (2.1) as the sum $x_1\varepsilon_1 + x_2\varepsilon_2 + x_3\varepsilon_3 + x_4\varepsilon_4$, where

$$\varepsilon_1 = \begin{pmatrix} 0 & -i \\ -i & 0 \end{pmatrix}, \quad \varepsilon_2 = \begin{pmatrix} 0 & 1 \\ -1 & 0 \end{pmatrix},$$

$$\varepsilon_3 = \begin{pmatrix} i & 0 \\ 0 & -i \end{pmatrix}, \quad \varepsilon_4 = \begin{pmatrix} 1 & 0 \\ 0 & 1 \end{pmatrix},$$

we can also identify $SU(2)$ with the quaternions of absolute value one.

In view of the simple form of the Fourier expansions (1.24) and (1.25), one would expect that the study of properties of central convolution operators is easier than that of the more general ones. In the case of $SU(2)$ one can essentially reduce the central case to the study of operators on the circle group. We see that this is plausible by making the following observations: It follows from elementary linear algebra considerations that if $u \in SU(2)$ then there exists $v \in SU(2)$ such that vuv^{-1} is a diagonal matrix of the form

$$e(\theta) = \begin{pmatrix} e^{i\theta} & 0 \\ 0 & e^{-i\theta} \end{pmatrix}.$$

Moreover, $e(-\theta) = u_0 e(\theta) u_0^{-1}$ if u_0 is the element of $SU(2)$,

$$u_0 = \begin{pmatrix} 0 & -1 \\ 1 & 0 \end{pmatrix}.$$

It follows that if f is a central function on $SU(2)$, then there exists an even periodic function f_0 on the real line (of period 2π) such that $f(u) = f_0(\theta)$, where $e^{i\theta}, e^{-i\theta}$ are the eigenvalues of u. Thus, there exists an obvious connection between central functions on $SU(2)$ and functions on T. In order to see how properties of f can be deduced from properties of f_0, we will need to consider the characters of $SU(2)$. Let us, therefore, examine briefly the representations of this group.

A standard way of obtaining all the irreducible representations of $SU(2)$ (up to equivalence) is the following. For each half-integer $l = 0, \frac{1}{2}, 1, \frac{3}{2}, 2, \ldots,$ let $\mathcal{P}^{2l}$ be the vector space of homogeneous polynomials in $z = (z_1, z_2) \in \mathbf{C}^2$ of degree $2l$. The $2l + 1$ polynomials $p_i(z) = z_1^i z_2^{2l-i}$, $i = 0, 1, \ldots, 2l$, obviously form a basis for this space and $\mathcal{P}^{2l}$ can be considered to be a Hilbert space if we impose the conditions that $\{p_0, p_1, \ldots, p_{2l}\}$ be an orthogonal basis and the norm of p_i is $1/\sqrt{\binom{2l}{i}}$, $i =$

0, 1, . . . , $2l$.* For each $u \in SU(2)$, we let

$$\left(S_u^{2l}p\right)(z) = p(zu), \tag{2.2}$$

whenever $p \in \mathcal{P}^{2l}$ and zu is the element of $\mathbf{C}^2$ obtained by multiplying the row vector $z = (z_1, z_2)$ on the right by the 2×2 matrix u. Then the mapping $S^{2l} : u \rightarrow S_u^{2l}$ is easily seen to be a representation of $SU(2)$ and it is not hard to show that it is irreducible. We can now calculate the characters associated with S^{2l} for each half-integer l.

We let t_{jk}^l, $0 \leqslant j, k \leqslant 2l$, denote the generalized trigonometric polynomials associated with the representation S^{2l} and the basis $\{p_0, p_1, \ldots, p_{2l}\}$. Since the characters are central functions, it suffices to find their values on the diagonal matrices $e(\theta)$. Since

$$\begin{aligned}\left(S_{e(\theta)}^{2l}p_j\right)(z) &= p_j(ze(\theta))\\ &= \left(e^{i\theta}z_1\right)^j\left(e^{-i\theta}z_2\right)^{2l-j}\\ &= e^{-2i(l-j)\theta}p_j(z),\end{aligned}$$

it follows that $t_{jj}^l(e(\theta)) = e^{-2i(l-j)\theta}$ for $0 \leqslant j \leqslant 2l$. Thus,

$$\begin{aligned}\chi^l(e(\theta)) &= \sum_{j=0}^{2l} e^{-2i(l-j)\theta}\\ &= e^{-2il\theta}\frac{1-\left(e^{2i\theta}\right)^{2l+1}}{1-e^{2i\theta}}\end{aligned}$$

and we obtain

$$\begin{aligned}\chi^l(e(\theta)) &= \chi_0^l(\theta)\\ &= \frac{\sin(2l+1)\theta}{\sin\theta}.\end{aligned} \tag{2.3}$$

---

*The symbol $\binom{2l}{i}$ denotes the binomial coefficient $(2l)!/i!(2l-i)!$. There are good reasons for introducing this normalization, but they are of a technical nature. We will not present them here.

Observe that $\chi_0^0(\theta) \equiv 1$, $\chi_0^{1/2}(\theta) = 2\cos\theta$ and, in general,

$$\tfrac{1}{2}\left(\chi_0^{l+1}(\theta) - \chi_0^l(\theta)\right) = \cos(2l+2)\theta$$

for $l = 0, \frac{1}{2}, 1, \frac{3}{2}, \ldots$. Hence, the space generated by the functions χ_0^l contains *all* the functions, $\cos k\theta$, $k = 0, 1, 2, \ldots$. It follows that the uniform closure (on $[0, 2\pi]$) of the span of the functions χ_0^l contains all even, continuous periodic functions. But we have seen that if f is central, then f_0 is even and periodic. From this we see that the functions χ^l are all the characters arising from irreducible representations of $SU(2)$.

Using formula (2.3), it is easy to express the Haar integral of a central function in terms of a Lebesgue integral on $[-\pi, \pi]$. Indeed, if f is central, we have

$$\int_{SU(2)} f(u)\,du = \frac{1}{\pi}\int_{-\pi}^{\pi} f_0(\theta)\sin^2\theta\,d\theta. \tag{2.4}$$

Since $\{\chi^l\}$, $l = 0, \frac{1}{2}, 1, \frac{3}{2}, \ldots,$ is an orthonormal family (see(1.11)), it suffices to verify (2.4) for the characters (for then this equality would hold for the finite linear combinations of characters and limiting arguments easily extend it to $L^2(SU(2))$). But, as we have seen, $\chi^0(u) \equiv 1$ and, thus, by orthonormality,

$$\int_{SU(2)} \chi^l(u)\,du = \int_G \chi^l(u)\overline{\chi^0(u)}\,du$$

$$= \begin{cases} 1 & \text{if} \quad l = 0, \\ 0 & \text{if} \quad l \neq 0. \end{cases}$$

On the other hand,

$$\frac{1}{\pi}\int_{-\pi}^{\pi} \chi_0^l(\theta)\sin^2\theta\,d\theta = \frac{1}{\pi}\int_{-\pi}^{\pi} \frac{\sin(2l+1)\theta}{\sin\theta}\sin^2\theta\,d\theta$$

$$= \begin{cases} 1 & \text{if} \quad l = 0, \\ 0 & \text{if} \quad l \neq 0, \end{cases}$$

and equality (2.4) follows.

These facts are all we need to indicate how properties of a central generalized convolution operator on $SU(2)$ can be studied by examining a related convolution operator on T. For simplicity, let us consider the convolution operator A_g, where g is a central function in $L^1(SU(2))$; that is,

$$(A_g f)(u) = \int_{SU(2)} g(v) f(uv^{-1})\, dv$$
$$= \int_{SU(2)} g(v) f(v^{-1}u)\, dv \tag{2.5}$$

whenever $f \in L^1(SU(2))$. We first make the observation that if h is an integrable function, then, by the invariance of Haar measure under translations,

$$\tilde{h}(w) = \int_{SU(2)} h(vwv^{-1})\, dv$$

defines a central function $\tilde{h}$ having the same integral as h. Making use of this fact, we have $g(w) = \int g(v^{-1}wv)dv$ by centrality of g, so by (2.4), we can rewrite (2.5) in the form

$$(A_g f)(u) = \frac{1}{\pi} \int_{SU(2)} \left\{ \int_{-\pi}^{\pi} \sin^2\theta\, g_0(\theta) f\big(uve(-\theta)v^{-1}\big)\, d\theta \right\} dv. \tag{2.6}$$

Let $k(\theta) = (1/\pi) \sin^2\theta\, g_0(\theta)$. The assumed integrability of g and (2.4) imply $k \in L^1(T)$. Denote by $N_p(k)$ the norm of the convolution operator induced by k on $L^p(T)$,* where $p \geqslant 1$. We shall now show that A_g *maps* $L^p(SU(2))$ *boundedly into itself with an operator norm not greater than* $N_p(k)$.

We want to estimate

$$\left(\int_{SU(2)} |(A_g f)(u)|^p\, du \right)^{1/p}.$$

---

*It is well known (and easy to prove) that $\|k * f\|_p \leqslant \|k\|_1 \|f\|_p$ for all $f \in L^p(T)$, $1 \leqslant p$. Thus, $N_p(k) \leqslant \|k\|_1 < \infty$.

By (2.6), this norm equals

$$\left[\int_{SU(2)}\left|\int_{SU(2)}\left\{\int_{-\pi}^{\pi}k(\theta)f\left(uve(-\theta)v^{-1}\right)d\theta\right\}dv\right|^{p}du\right]^{1/p}.$$

By Minkowski's integral inequality, we see that this expression is dominated by

$$\int_{SU(2)}\left[\int_{SU(2)}\left|\int_{-\pi}^{\pi}k(\theta)f\left(uve(-\theta)v^{-1}\right)d\theta\right|^{p}du\right]^{1/p}dv.$$

Because of the invariance of Haar measure the integral within the parenthesis is unchanged if we multiply u, on the right, by $ve(\psi)$ instead of just v. Doing so and averaging over all $\psi\in[-\pi,\pi]$ we obtain

$$\int_{SU(2)}\left|\int_{-\pi}^{\pi}k(\theta)f\left(uve(-\theta)v^{-1}\right)d\theta\right|^{p}du$$

$$=\int_{SU(2)}\left\{\frac{1}{2\pi}\int_{-\pi}^{\pi}\left|\int_{-\pi}^{\pi}k(\theta)f\left(uve(\psi-\theta)v^{-1}\right)d\theta\right|^{p}d\psi\right\}du.$$

But the expression in curly brackets is the pth power of the L^p norm of the convolution of k with the function h having values $h(\varphi)=f(uve(\varphi)v^{-1})$. Thus, the last integral is no larger than

$$\frac{1}{2\pi}(N_p(k))^p\int_{SU(2)}\left\{\int_{-\pi}^{\pi}\left|f\left(uve(\varphi)v^{-1}\right)\right|^{p}d\varphi\right\}du. \tag{2.7}$$

Interchanging the order of integration and again using the invariance of Haar measure we see that (2.7) is dominated by

$$(N_p(k))^p\int_{SU(2)}|f(u)|^p du.$$

Collecting these estimates we obtain the announced result:

$$\left(\int_{SU(2)} |(A_g f)(u)|^p du\right)^{1/p} \leq N_p(k)\left(\int_{SU(2)} |f(u)|^p du\right)^{1/p}.$$

One can extend this inequality to various central generalized convolution operators. This is usually done by approximating these operators by ones that we just considered. The crucial property that allows us to do this is the control we have of the operator norms involved; specifically, we showed that the same norm can be used when we transfer, by this method, operators associated with the circle to operators associated with $SU(2)$. We shall now give a general result that illustrates this general *transference* method.

There exist simple sufficient conditions on the Fourier coefficients of a generalized convolution operator $A \sim \sum_{k=-\infty}^{\infty} a_k e_k$ for it to be a bounded operator on $L^p(T)$ for $1 < p < \infty$. These conditions are that both the sequence (a_k), $k = 0, \pm 1, \pm 2, \ldots$, and the sequence

$$(b_k) = \left(\sum_{2^k \leq |j| < 2^{k+1}} |a_{j+1} - a_j|\right)$$

$k = 0, 1, 2, \ldots$, be bounded. Moreover, the norm of the operator A on $L^p(T)$ depends only on the least upper bounds of these two sequences. This is a celebrated result known as the Marcinkiewicz multiplier theorem. We shall show how it can be transferred to $SU(2)$.

Suppose

$$B \sim \sum_{2l=0}^{\infty} b_l \chi^l$$

is a central generalized convolution operator on $SU(2)$. The method we just described essentially tells us that B will map $L^p(SU(2))$ boundedly into itself if the generalized convolution operator

$$A \sim \sum_{2l=0}^{\infty} b_l \sin^2\theta \, \chi_0^l(\theta)$$

maps $L^p(T)$ boundedly into itself. But, by (2.3),

$$\begin{aligned} \sin^2\theta \, \chi_0^l(\theta) &= \sin\theta \sin(2l+1)\theta \\ &= \tfrac{1}{2}(\cos 2l\theta - \cos(2l+2)\theta). \end{aligned}$$

Thus,

$$A \sim \tfrac{1}{2} \sum_{k=0}^{\infty} (\Delta b_k) \cos k\theta, \tag{2.8}$$

where $\Delta b_0 = b_0$, $\Delta b_1 = b_{1/2}$ and $\Delta b_k = b_{k/2} - b_{(k-2)/2}$ if $k = 2, 3, 4, \ldots$. Applying the Marcinkiewicz theorem to the operator A having Fourier expansion (2.8), we obtain the following multiplier theorem for $SU(2)$:

If $B \to \Sigma_{2l=0}^{\infty} b_l \chi^l$ is a central generalized convolution operator on $SU(2)$ such that the sequences (Δb_k) and $\Sigma_{2^k \leqslant j < 2^{k+1}} |\Delta b_{j+1} - \Delta b_j|$, $k = 0, 1, 2, \ldots$, are bounded, then B maps $L^p(SU(2))$, $1 < p < \infty$, boundedly into itself.

Without going into any details, we would like to point out that this method of transferring boundedness of operators, from T to $SU(2)$, can be extended to other properties of operators or sequences of operators. In particular, from the behaviour of the sequence of partial sums of Fourier series one can obtain results involving summability of Fourier expansions of functions in $L^p(SU(2))$. In view of the close relationship between the characters χ^l and the Dirichlet kernels (observe that, for integral l, $\chi_0^l(\pi\theta) = D_l(\theta)$), it should not be surprising that the role played by the partial sums of Fourier series is assumed by the Cesáro means of Fourier expansions of functions on $SU(2)$.

We chose these examples of results in harmonic analysis on the group $SU(2)$ since they illustrate what can be done more generally on an important class of groups, the class of *compact Lie groups*. It is not really necessary to define the general concept of a Lie group in our case because one can classify the *connected compact* Lie groups (up to "local isomorphisms") in terms of a relatively small collection of basic matrix groups and their products. For example, the *n-torus* $T^n = T \times T \times \cdots \times T$ (n factors), $n = 1, 2, 3, \ldots$, is the most general compact connected *abelian* Lie group. The *rotation groups* $SO(n)$ consisting of the $n \times n$ orthogonal matrices of determinant 1, the *special unitary groups* $SU(n)$ made up by the $n \times n$ unitary matrices of determinant 1, the *symplectic groups* $Sp(n)$, which are defined in complete analogy with the last two examples by making use of the quaternions instead of the real and complex fields, and five exceptional groups make up this collection of matrix groups.

For the compact Lie groups one can find a homomorphic image of the n-torus T^n, for an appropriate n, on which the central functions are completely determined. The method we described above for studying central convolution operators on $SU(2)$ can be extended to the general compact Lie group G. Thus, results associated with the n-torus can be transferred to G.

Up to this point we concentrated mostly on "central" results. The methods we used can be employed to obtain certain non-central multiplier theorems. In order to study the general multiplier operators (see (1.18) and (1.19)), however, other techniques have to be introduced. By these means one can obtain multiplier theorems giving sufficient conditions on the matrices $m(\lambda)$ to insure boundedness on $L^p(G)$, $1 < p < \infty$. In the case of $SU(2)$, such results automatically give us theorems about the development of functions in series involving Jacobi polynomials. This is the case because one can show that the generalized trigonometric polynomials t^l_{jk} are intimately connected with these classical polynomials $P_n^{\alpha,\beta}$. The relationships we are discussing are of the following type:

$$t^l_{jk}(a(\theta)) = c^l_{j,k}\left(\cos\frac{\theta}{2}\right)^{\beta}\left(\sin\frac{\theta}{2}\right)^{\alpha} P_n^{\alpha,\beta}(\cos\theta),$$

when $\alpha = j - k \geqslant 0, \beta = 2l + j + k \geqslant 0, n = l - j$ and

$$a(\partial) = \begin{pmatrix} \cos(\theta/2) & i\sin(\theta/2) \\ i\sin(\theta/2) & \cos(\theta/2) \end{pmatrix}.$$

Similar equalities hold for the other indices l, j, k.

We have not discussed how one can construct the irreducible representations of compact groups. In the case of the matrix groups, one can give a very explicit algebraic construction (in terms of tensor products). The orthogonality relations of the generalized trigonometric polynomials and the nature of this algebraic process explain why the various classical functions we mentioned above appear in the study of the representations of these groups. In fact, we obtain a modern method for studying these classical objects. Many of the classical formulae involving special functions are better understood from this viewpoint and their derivation is usually much less complicated. In some cases one even obtains very useful relations that seem not to have been discovered before.

3. GENERAL OBSERVATIONS AND A VERY SHORT GUIDE TO THE LITERATURE

The structure of the compact groups has been studied in great detail by many mathematicians and physicists. The particular form of the characters and the Haar integral for central functions were discovered by H. Weyl. The book of Adams [1] presents a readable account of these subjects as well as representation theory for compact Lie groups. Stein's book [6] deals with many topics in harmonic analysis on compact groups and includes the study of multipliers from another point of view. A very detailed study of matrix groups and the associated theory of special functions can be found in Vilenkin's book [7]. The harmonic analysis on $SU(2)$ and some of its applications are presented in [2]. In these lecture notes we prove a multiplier theorem for general convolution operators. The method in this proof, based on an extension of the Calderón-Zygmund singular integral theory, shows that, as is the case in the classical theory, one can study the properties of functions and, more generally, convolution operators by examining "regularity conditions" for their matricial Fourier coefficients. These regularity conditions are given by difference operators involving matrices of different sizes. These are studied in great detail when G is $SU(2)$ and $SO(3)$. It is of basic interest to continue this detailed study for the higher dimensional compact Lie groups.

At the end of the last section we mentioned the fact that many useful relations among special functions can be derived by group theoretic methods. An example of one such equality that appears to have escaped previous attention is given on page 65 of [2].

The extension to the general compact group of the transference method described here for $SU(2)$ can be found in [4]. In this paper we obtain a general central multiplier theorem for compact Lie groups. Besides Stein's book, which we mentioned above, another method for obtaining a multiplier theorem for such groups is given by N. J. Weiss [8].

Not only did we not define the concept of a Lie group, but we did not mention its Lie algebra. This is just one of many topics that have been omitted from our lectures due to lack of time. The omission of a subject on our part, however, should not be grounds for concluding that we do not think it is important. In particular, the Lie algebra of a compact group is an important notion for many aspects of harmonic analysis. An example of its use in the study of "complex methods" and, in particular, the study of H^p spaces is given in [3]. We refer the reader to [5] for a treatment of Lie algebras in connection with compact groups.

REFERENCES

1. Adams, J. Frank, *Lectures on Lie Groups*, W. A. Benjamin, Inc., New York, 1969.

2. Coifman, R. R., and Weiss, Guido, *Analyse Harmonique Non-Commutative sur Certains Espaces Homogènes*, vol. 242, Springer-Verlag, Berlin, 1971.

3. ——, "Invariant Systems of Conjugate Harmonic Functions Associated with Compact Lie Groups," *Studia Math.*, T.XLIV., (1972), 301–308.

4. ——, "Central Multiplier Theorems for Compact Lie Groups," *Bull. Amer. Math. Soc.*, **80** No. 1, (1974), 124–126.

5. Hausner, Melvin, and Schwartz, Jacob T., *Lie Groups; Lie Algebras*, Gordon and Breach, New York, 1968.

6. Stein, E. M., "Topics in Harmonic Analysis," *Annals of Mathematics Study* (63), Princeton University Press (1970).

7. Vilenkin, N. J., "Special Functions and the Theory of Group Representations," *AMS Translations*, **22** (1968).

8. N. J. Weiss, "L^p Estimates for Bi-invariant Operators on Compact Lie Groups", *Amer. J. Math.*, **94** (1972), 103–118.

HARMONIC ANALYSIS AND GROUP REPRESENTATIONS

Paul J. Sally, Jr.

1. INTRODUCTION

The study of harmonic analysis on a general locally compact group involves many factors which are not present in harmonic analysis on abelian groups and compact groups. The goals are similar to those of the abelian and compact theories; that is, the analysis of certain function spaces on the group. In particular, in this paper, we shall discuss analogues of Fourier inversion for spaces of "smooth" functions and Plancherel's theorem for square-integrable functions. Basic to all of this is a discussion of Haar measure, representation theory and character theory. We treat these topics in the first few sections of the paper. We shall attempt to demonstrate the nature of the subject by stating theorems, mostly without proof, and working out some examples in detail. References are provided for the proofs of stated theorems, and a brief guide to the literature is given in the last section. It should be remarked that the theory of harmonic analysis on a general locally compact group is in a very primitive state. Only by restricting our attention to specific classes of groups with a sufficiently rich structure is it possible to obtain a satisfactory theory.

2. LOCALLY COMPACT GROUPS AND HAAR MEASURE

Let G be a group which is also a Hausdorff topological space. We say that G is a *topological group* if the map from $G \times G$ (endowed with the product topology) into G defined by $(x, y) \mapsto xy^{-1}$ is continuous. A topological group G is a *locally compact group* if G is locally compact as a topological space. A discussion of the elementary properties of topological and, in particular, locally compact groups may be found in Pontryagin [22], Ch. III, or Hewitt and Ross [12], v. I, Ch. II. *In this paper, we shall work only with separable locally compact groups.*

The most familiar examples of locally compact groups are the following—all taken with the usual topology:

R— the additive group of real numbers;
R^*— the multiplicative group of non-zero real numbers;
R^*_+—the multiplicative group of positive real numbers;
C— the additive group of complex numbers;
C^*— the multiplicative group of non-zero complex numbers;
T— the multiplicative group of complex numbers with absolute value equal to one.

Of course, the above groups are all abelian.

Before presenting further examples, we gather a few useful facts in the form of a proposition, the proof of which may be found in [22], Ch. III.

PROPOSITION 2.1: *Suppose that G is a locally compact group and that H is a closed subgroup of G. Then*

(a) *H is a locally compact group in the relative topology from G;*
(b) *if H is a normal subgroup of G, then G/H is a locally compact group;*

(c) *if G' is a locally compact group, then $G \times G'$ with the product topology is a locally compact group.*

Now, let F denote either R or C. We denote by $M_2(F)$ the ring of two by two matrices over F. As an additive group, $M_2(F)$ is isomorphic to $F^4(= F \times F \times F \times F)$. It follows from Proposition 2.1(c) that $M_2(F)$ is a locally compact group in the product topology. Let $GL_2(F)$ denote the collection of all non-singular matrices in $M_2(F)$, that is, all two by two matrices over F with non-zero determinant. Then, $GL_2(F)$ is a group (under multiplication of matrices), and, moreover, $GL_2(F)$ is an open, dense subset of $M_2(F)$. It follows immediately that $GL_2(F)$ is a locally compact group in the relative topology from $M_2(F)$. $GL_2(F)$ is called the two by two *general linear group* over F. The locally compact groups which we use for examples in our present development of harmonic analysis are

$$SL_2(F) = \left\{ \begin{pmatrix} \alpha & \beta \\ \gamma & \delta \end{pmatrix} \in GL_2(F) : \alpha\delta - \beta\gamma = 1 \right\},$$

and

$$C_2 = \left\{ \begin{pmatrix} 1 & z \\ 0 & e^{i\theta} \end{pmatrix} \in GL_2(C) : z \in C \quad \text{and} \quad e^{i\theta} \in T \right\}.$$

These are closed subgroups of $GL_2(F)$ and $GL_2(C)$ respectively, and hence, by Proposition 2.1(a), they are locally compact groups in the relative topology. The group $SL_2(F)$ is called the two by two *special linear group* over F. The group C_2 is called the *group of rigid motions* of the plane since the elements of C_2 may be regarded as distance preserving transformations of the complex plane. Thus, if we write a complex number w as a row matrix $(1 \;\; w)$, we have

$$(1 \quad w)\begin{pmatrix} 1 & z \\ 0 & e^{i\theta} \end{pmatrix} = (1 \quad z + we^{i\theta}).$$

This represents a rotation of w through an angle θ followed by a translation through z.

The group C_2 is a particular example of a class of groups called semi-direct products. In general, a locally compact group G is a *semi-direct product* of closed subgroups H and N if (1) $G = HN$, (2) $H \cap N = \{e\}$, where e is the identity in G, and (3) N is a normal subgroup of G. For C_2, we take

$$H = \left\{ \begin{pmatrix} 1 & 0 \\ 0 & e^{i\theta} \end{pmatrix} : e^{i\theta} \in T \right\}$$

and

$$N = \left\{ \begin{pmatrix} 1 & z \\ 0 & 1 \end{pmatrix} : z \in C \right\}.$$

We now turn to a discussion of Haar measure. Haar measure on a locally compact group G is a natural generalization of Lebesgue measure on R (or R^n) in that it provides us with a positive Borel measure on G which is nicely related to the group operation (that is, translation). Of course, we must take into account the possibility that G may be non-abelian.

PROPOSITION 2.2: *If G is a locally compact group, then there exists a positive Borel measure μ defined on G which is invariant under right translation. Thus, if E is a μ-measurable set,*

$$\mu(Ex) = \mu(E) \qquad \text{for all} \qquad x \in G. \tag{2.3}$$

The measure μ is unique up to multiplication by a positive constant.

We call μ a *right Haar measure* on G. A complete treatment of Haar measure may be found in the books of Nachbin [19] and Weil [35]. In this paper, we restrict ourselves to a few elementary facts necessary for our development of representation theory.

Let $C_c(G)$ denote the space of complex-valued, compactly supported, continuous functions on G. We write symbolically $\mu = d_r x$. Then, for $f \in C_c(G)$, we have

$$\int_G f(xy)\, d_r x = \int_G f(x)\, d_r x, \qquad y \in G. \tag{2.4}$$

A *left Haar measure* $d_l x$ on G is defined in an analogous way. In this case, for $f \in C_c(G)$, we have

$$\int_G f(yx)\, d_l x = \int_G f(x)\, d_l x, \qquad y \in G. \tag{2.5}$$

As in the case of right Haar measure, left Haar measure is unique up to multiplication by a positive constant. If we fix the right (or left) Haar measure of some set of positive measure, then the constant mentioned above is fixed and we obtain a *normalization* of right (or left) Haar measure.

The properties that essentially characterize right Haar measure along with (2.4) are

(M1) non-empty open sets have positive measure;
(M2) compact sets have finite measure.

Left and right Haar measure on a locally compact group G are related in a nice way. To see this, let $d_r x$ be a right Haar measure on G. Then, for a fixed element $y \in G$, the map $f \mapsto \int_G f(yx)\, d_r x$, $f \in C_c(G)$, also defines a right Haar measure on G which must differ from $d_r x$ by a positive constant. Thus, there exists a positive real number $\Delta_G(y)$ such that

$$\int_G f(yx)\, d_r x = \Delta_G(y) \int_G f(x)\, d_r x, \qquad f \in C_c(G). \tag{2.6}$$

It is not difficult to see that Δ_G is a continuous homomorphism of G into R_+^*. We shall call Δ_G the *modular function* on G.

Now, given $d_r x$, we set $d_l x = \Delta_G(x)\, d_r x$. Then, for $f \in C_c(G)$,

$$\int_G f(yx) \Delta_G(x)\, d_r x = \int_G f(x) \Delta_G(x)\, d_r x. \tag{2.7}$$

Thus, $d_l x = \Delta_G(x)\, d_r x$ is a left Haar measure on G. In the particular case when $\Delta_G = 1$, that is, when every right Haar measure is also a left Haar measure, the group G is called *unimodular*.

Remarks 2.8. (i) Abelian groups are obviously unimodular.

(ii) Compact groups are unimodular since the image of Δ_G is a compact subgroup of R_+^*. We note that a locally compact group is compact if and only if the total volume of the group is finite.

(iii) Discrete groups (that is, any group with the discrete topology) are unimodular since all points have the same positive measure (points are open sets).

Examples 2.9. (i) Let $G = R$. Then ordinary Lebesgue measure is a left and right Haar measure on G. The normalization is chosen so that $\int_0^1 dx = 1$. Similarly, $dx/|x|$ is a left and right Haar measure on R^*. Here, the normalization is

$$\int_1^e \frac{dx}{|x|} = 1.$$

(ii) For $G = C$, we take $dz = dx\, dy$ ($z = x + iy$). This is just the product measure on $R \times R$ which is isomorphic to C. On C^*, we take $dz/|z|^2$ for a Haar measure. Note that a normalization is already built into these measures since they are constructed from the measure dx on R normalized as above. For convenience, let us define

$$\operatorname{mod}(\alpha) = \begin{cases} |\alpha|, & \alpha \in R^* \\ |\alpha|^2, & \alpha \in C^* \end{cases}.$$

Then, if F denotes either R or C, a Haar measure on F^* is given by $d\alpha/\operatorname{mod}(\alpha)$, and, for any $\gamma \in F^*$, we have $d(\gamma\alpha) = \operatorname{mod}(\gamma)\, d\alpha$.

(iii) Let

$$B = \left\{ \begin{pmatrix} \alpha & \beta \\ 0 & \alpha^{-1} \end{pmatrix} : \alpha \in F^*, \quad \beta \in F \right\}.$$

Then B is a closed subgroup of $SL_2(F)$ and hence is a locally compact group. A right Haar measure on B is given by $d_r b = d\alpha\, d\beta$, and a left Haar measure by $d_l b = d\alpha\, d\beta/\operatorname{mod}(\alpha)^2$.

Thus, if

$$b' = \begin{pmatrix} \gamma & \delta \\ 0 & \gamma^{-1} \end{pmatrix} \in B,$$

it can be shown by a simple change of variables that, for $f \in C_c(B)$,

$$\begin{aligned}\int_B f(bb')\, d_r b &= \int_F \int_{F^*} f\left[\begin{pmatrix} \alpha & \beta \\ 0 & \alpha^{-1} \end{pmatrix}\begin{pmatrix} \gamma & \delta \\ 0 & \gamma^{-1} \end{pmatrix}\right] d\alpha\, d\beta \\ &= \int_F \int_{F^*} f\left[\begin{pmatrix} \alpha & \beta \\ 0 & \alpha^{-1} \end{pmatrix}\right] d\alpha\, d\beta \\ &= \int_B f(b)\, d_r b.\end{aligned}$$

Similarly,

$$\int_B f(b'b)\, d_l b = \int_B f(b)\, d_l b, \qquad \text{for} \qquad f \in C_c(B).$$

We conclude that B is not unimodular and that

$$\Delta_B\left[\begin{pmatrix} \alpha & \beta \\ 0 & \alpha^{-1} \end{pmatrix}\right] = \operatorname{mod}(\alpha)^{-2}.$$

(iv) The group $G = SL_2(F)$ is unimodular. If $g = \begin{pmatrix} \alpha & \beta \\ \gamma & \delta \end{pmatrix} \in G$, then $dg = d\beta\, d\gamma\, d\delta/\operatorname{mod}(\delta)$ is a (left and right) Haar measure on G. This may be verified by using the fact that G is generated by $\begin{pmatrix} 0 & 1 \\ -1 & 0 \end{pmatrix}$ and elements of the form $\begin{pmatrix} 1 & \beta \\ 0 & 1 \end{pmatrix}$, $\beta \in F$. Note that this particular Haar measure on G is not defined on the set of measure zero where $\delta = 0$.

(v) The group $G = C_2$ is unimodular, and if $g = \begin{pmatrix} 1 & z \\ 0 & e^{i\theta} \end{pmatrix} \in G$, then $dg = (1/2\pi)\, d\theta\, dz$ is a Haar measure on G.

3. REPRESENTATION THEORY—DEFINITIONS AND EXAMPLES

Let G be a locally compact group and $\mathcal{H}$ a (complex) Hilbert space. We denote by $\mathcal{L}(\mathcal{H})$ the algebra of bounded linear operators on $\mathcal{H}$. A *representation* of G on $\mathcal{H}$ is a map $T: G \to \mathcal{L}(\mathcal{H})$ such that

(R1) $T(xy) = T(x)T(y)$ for all $x, y \in G$:

(R2) $T(e) = I$, where e is the identity element in G and I is the identity operator on $\mathcal{H}$;

(R3) T is strongly continuous, that is, for each $v \in \mathcal{H}$, the map $x \mapsto T(x)v$ is continuous from G to $\mathcal{H}$.

We shall write $(T, \mathcal{H})$ for the representation. If $\mathcal{H}_0$ is a subspace of $\mathcal{H}$, we say that $\mathcal{H}_0$ is *invariant* (under T) if $T(x)\mathcal{H}_0 \subset \mathcal{H}_0$ for all $x \in G$. A representation $(T, \mathcal{H})$ of G is *irreducible* if the only closed subspaces of $\mathcal{H}$ which are invariant under T are $\{0\}$ and $\mathcal{H}$. We remark that the concept of irreducibility is related to *closed* subspaces. A representation is *algebraically irreducible* if there are no non-trivial invariant subspaces, closed or otherwise. Of course, the notions of irreducibility and algebraic irreducibility coincide if the dimension of $\mathcal{H}$ is finite.

If $(T, \mathcal{H})$ is a representation of a locally compact group G and $\mathcal{H}$ is an infinite-dimensional space, we say that T is an *infinite-dimensional representation*. On the other hand, if $\mathcal{H}$ has finite dimension, we say that T is *finite-dimensional*, and we denote the dimension of $\mathcal{H}$ by d_T. The dimension d_T is called the *degree* of T.

A representation $(T, \mathcal{H})$ is a *unitary representation* if $T(x)$ is a unitary operator for all $x \in G$. In this case, we may say that T is a homomorphism of G into the group $\mathcal{U}(\mathcal{H})$ of unitary operators on $\mathcal{H}$ which is continuous in the sense of (R3). The irreducible unitary representations of a locally compact group are the basic building blocks of harmonic analysis on the group. They are the analogues of the characters of a locally compact abelian group. As a matter of fact, if G is a locally compact abelian group and χ is a character on G [25], we can define a unitary representation T of G on the one-dimensional Hilbert space C by

$$T(x)z = \chi(x)z, \qquad x \in G, z \in C. \tag{3.1}$$

Obviously, this representation is irreducible. As we shall see below, every irreducible unitary representation of a locally compact abelian group has this form. Thus, the irreducible unitary representations of a locally compact abelian group are all one-dimensional.

As mentioned in the article of Weiss, every irreducible unitary representation of a compact group is finite-dimensional. This is decidedly not the case for non-compact, non-abelian groups, and that's where the fun begins. One problem which we face is that of constructing irreducible unitary representations of a given group. In the next section, we illustrate the most effective technique available, that is, the construction of induced representations.

At this point, we observe that a representation of a locally compact group G involves a Hilbert space which, *a priori*, has nothing to do with G. Thus, if we have two unitary representations $(T_1, \mathcal{H}_1)$ and $(T_2, \mathcal{H}_2)$ of G, we may ask if T_1 and T_2 are the same representation in some sense. This leads to the notion of unitary equivalence.

We say that $(T_1, \mathcal{H}_1)$ and $(T_2, \mathcal{H}_2)$ are *unitarily equivalent* if there exists a unitary operator $A: \mathcal{H}_1 \to \mathcal{H}_2$ such that

$$AT_1(x) = T_2(x)A \qquad \text{for all} \qquad x \in G. \tag{3.2}$$

Such an operator A is called an *intertwining operator* for T_1 and T_2. Clearly, unitary equivalence is an equivalence relation on the collection of all irreducible unitary representations of G. The set of equivalence classes of irreducible unitary representations of G is called the *unitary dual* of G and is denoted by $\hat{G}$. For a non-compact, non-abelian group G, the determination of $\hat{G}$ may be a very difficult problem. However, it may not be necessary to determine $\hat{G}$ completely in order to derive a Fourier inversion formula or Plancherel's formula for G. This is illustrated in the later sections of the paper.

Before presenting a few examples, we state the basic irreducibility criterion for unitary representations of a locally compact group.

Irreducibility criterion. ([2], p. 599, [10], Ch. I) (3.3)

Let $(T, \mathcal{H})$ be a unitary representation of a locally compact group G. Then T is irreducible if and only if every $A \in \mathcal{L}(\mathcal{H})$ which satisfies the condition $AT(x) = T(x)A$ for all $x \in G$ has the form $A = cI$ where $c \in C$ and I is the identity operator on $\mathcal{H}$.

As an immediate corollary, we have the fact that every irreducible unitary representation of a locally compact abelian group is one-dimensional. To see this, suppose that $(T, \mathcal{H})$ is an irreducible unitary representation of a locally compact abelian group G and fix $y \in G$. Then

$$T(y)T(x) = T(yx) = T(xy) = T(x)T(y) \qquad \text{for all} \qquad x \in G.$$

It follows from (3.3) that $T(y) = \chi(y)I$ where $\chi(y) \in C$. Since T is irreducible, $\mathcal{H}$ must be one-dimensional. Moreover, the map $y \mapsto \chi(y)$, $y \in G$, defines a character of G as we indicated previously. Using (3.1) and this last remark, we conclude that, for a locally compact abelian group G, the unitary dual $\hat{G}$ defined above may be identified with the character group of G.

Examples 3.4. (i) Let G be a locally compact group and let $\mathcal{H} = C$. For $x \in G$, set $T(x) = I$ where I is the identity operator on C. Then T is an irreducible unitary representation of G on C which is called the *trivial representation* of G.

(ii) Let G be a locally compact group and assume that G is unimodular. Let $\mathcal{H} = L^2(G)$, the Hilbert space of equivalence classes of measurable functions which are square-integrable with respect to Haar measure. (We note that, if G is not unimodular, then the spaces of functions which are square-integrable with respect to right and left Haar measure respectively are not identical.) For $f \in \mathcal{H}$ and $x \in G$, we define

$$[R(x)f](y) = f(yx), \qquad y \in G. \tag{3.5}$$

Then $(R, \mathcal{H})$ is a unitary representation of G which is irreducible if and only if G has exactly one element. The representation R is called the *right regular representation* of G. The *left regular representation* (L, $\mathcal{H}$) of G is defined by

$$[L(x)f](y) = f(x^{-1}y), \qquad x, y \in G, \quad f \in \mathcal{H}. \tag{3.6}$$

The unitary representations R and L are unitarily equivalent via the map $A \in \mathfrak{U}(\mathcal{H})$ given by $[Af](x) = f(x^{-1})$, $f \in \mathcal{H}$, $x \in G$. The map A is unitary because, on a unimodular group, the map $x \mapsto x^{-1}$ is measure preserving. The representations R and L are finite-dimensional if and only if G is a finite group.

(iii) Let $G = C_2$, the motion group of the plane. For any integer n, define $\epsilon_n(e^{i\theta}) = e^{in\theta}$, $e^{i\theta} \in T$. Take $\mathcal{H} = C$ and set

$$T_n\begin{pmatrix} 1 & z \\ 0 & e^{i\theta} \end{pmatrix} w = \epsilon_n(e^{i\theta})w, \quad \begin{pmatrix} 1 & z \\ 0 & e^{i\theta} \end{pmatrix} \in G, \quad w \in \mathcal{H}. \tag{3.7}$$

Then, T_n is a unitary representation of G which is one-dimensional and hence irreducible. Note that the case $n = 0$ yields the trivial representation of G. The representations T_n do not appear in the Plancherel formula for G. All the remaining irreducible unitary representations of G are infinite-dimensional. These representations, which are constructed in the next section, are used to determine the Plancherel formula for G.

(iv) Let $G = SL_2(C)$ and let k be a non-negative integer. Denote by V_k the vector space of homogeneous polynomials in two variables over C having degree k. V_k is a vector space of degree $k+1$ over C, and the set $\{z_1^k, z_1^{k-1}z_2, \ldots, z_1z_2^{k-1}, z_2^k\}$ is a basis for V_k. For $g = \begin{pmatrix} \alpha & \beta \\ \gamma & \delta \end{pmatrix} \in G$ and $p \in V_k$, we set

$$[T_k(g)p](z_1, z_2) = p(\alpha z_1 + \gamma z_2, \beta z_1 + \delta z_2), \; (z_1, z_2) \in C^2. \tag{3.8}$$

With the obvious inner product on V_k, T_k is a representation of G on V_k of degree $k+1$. It is not difficult to prove that T_k is irreducible for all $k \geqslant 0$. However, for $k > 0$, there is no inner product on V_k with respect to which T_k is unitary (of course, $k = 0$ yields the trivial representation which is unitary). As a matter of fact, G has no finite-dimensional unitary representations other than the trivial representation ([12], v. I, p. 348). Thus, for the study of harmonic analysis on G it is necessary to consider infinite-dimensional representations.

4. CONSTRUCTION OF REPRESENTATIONS—INDUCED REPRESENTATIONS

Let $G = SL_2(F)$ where $F = R$ or C. We shall first write down a collection of infinite-dimensional unitary representations of G and then indicate how these representations arise in a natural way.

Let π be a character on the locally compact abelian group F^*, that is, $\pi \in (F^*)\hat{}$. Then, for $F = R$, π has the form

$$\pi(x) = \left(\frac{x}{|x|}\right)^h |x|^{is}, \qquad h = 0, 1, \qquad s \in R, \tag{4.1}$$

and, for $F = C$,

$$\pi(z) = \left(\frac{z}{|z|}\right)^n |z|^{is}, \qquad n \in Z, \qquad s \in R. \tag{4.2}$$

Here, Z denotes the ring of integers.

Take $\pi \in (F^*)\hat{}$, $f \in L^2(F)$, $g = \begin{pmatrix} \alpha & \beta \\ \gamma & \delta \end{pmatrix} \in G$, and define

$$\left[T_\pi\begin{pmatrix} \alpha & \beta \\ \gamma & \delta \end{pmatrix} f\right](v) = [\operatorname{mod}(\beta v + \delta)]^{-1}\pi(\beta v + \delta) f\left(\frac{\alpha v + \gamma}{\beta v + \delta}\right),$$
$$v \in F. \tag{4.3}$$

We remark that the above formula has no meaning when $\beta v + \delta = 0$, but, since the operator $T_\pi(g)$ is acting on $L^2(F)$, the formula holds only up to sets of measure zero in any case.

PROPOSITION 4.4: *For $\pi \in (F^*)\hat{}$, the map $g \mapsto T_\pi(g)$ defines a unitary representation of $SL_2(F)\hat{}$ on $L^2(F)$. This representation is irreducible with the single exception of $\pi(x) = \dfrac{x}{|x|} = \operatorname{sgn}(x)$ when $F = R$.*

The proof of this proposition is not difficult ([7], Ch. III and Ch. VII, [10], Ch. I). The reader should have no problem verifying (R1), (R2) and (R3). The collection of representations $\{T_\pi$:

$\pi \in (F^*)^\wedge\}$ is called the *principal series* of representations of $G = SL_2(F)$.

We now show that the representations T_π can be constructed from one-dimensional unitary representations of the closed subgroup B of G defined in Examples 2.9(iii). In this context, the representations T_π, $\pi \in (F^*)^\wedge$, are called *induced representations*.

Let A and N be the closed subgroups of G defined by

$$A = \left\{ \begin{pmatrix} \alpha & 0 \\ 0 & \alpha^{-1} \end{pmatrix} : \alpha \in F^* \right\}, \tag{4.5}$$

$$N = \left\{ \begin{pmatrix} 1 & \beta \\ 0 & 1 \end{pmatrix} : \beta \in F \right\}. \tag{4.6}$$

Then A is topologically isomorphic to F^*, and N is topologically isomorphic to F. (A topological isomorphism is an algebraic isomorphism which is also a homeomorphism.) The group B is the semi-direct product of A and N, N being the normal subgroup. Thus, any character on F^* yields a character on A, and, if $\pi \in (F^*)^\wedge$(or $\hat{A}$), we may extend π to B by the formula

$$\pi\begin{pmatrix} \alpha & \beta \\ 0 & \alpha^{-1} \end{pmatrix} = \pi(\alpha)^{-1}. \tag{4.7}$$

The map $b \mapsto \pi(b)$ defines a one-dimensional unitary representation of B in the manner of (3.1). We mention in passing that every finite-dimensional, irreducible unitary representation of B has the above form ([9], p. 549).

Let C_π be the linear space of continuous, complex-valued functions on G satisfying the condition

$$f(bg) = \Delta_B(b)^{-1/2}\pi(b)f(g), \quad b \in B, \quad g \in G. \tag{4.8}$$

For $g' \in G$ and $f \in C_\pi$, we define

$$[T^\pi(g')f](g) = f(gg'). \tag{4.9}$$

Then, T^π acts on the space C_π like the right regular representation and hence satisfies the algebraic condition (RI). It is possible to define an inner product on C_π so that the operators $T^\pi(g), g \in G$, are isometries on C_π with respect to this inner product. If C_π is completed to a Hilbert space $\mathcal{H}_\pi$, then T^π extends to a unitary representation of G on $\mathcal{H}_\pi$. A complete development of the procedure outlined above may be found in [10], and a brief summary in [26].

We shall not elaborate further on the analytic aspects of the above construction. Our purpose here is to show how the formula (4.3) is obtained from (4.8) and (4.9). To this end, we observe that G can be decomposed as follows. If $g = \begin{pmatrix} \alpha & \beta \\ \gamma & \delta \end{pmatrix} \in G$, then

$$g = \begin{cases} \begin{pmatrix} \delta^{-1} & \beta \\ 0 & \delta \end{pmatrix}\begin{pmatrix} 1 & 0 \\ \delta^{-1}\gamma & 1 \end{pmatrix}, & \text{if } \delta \neq 0, \\ \begin{pmatrix} \beta & -\alpha \\ 0 & \beta^{-1} \end{pmatrix}\begin{pmatrix} 0 & 1 \\ -1 & 0 \end{pmatrix}, & \text{if } \delta = 0. \end{cases} \tag{4.10}$$

Thus, if $p = \begin{pmatrix} 0 & 1 \\ -1 & 0 \end{pmatrix}$ and $V = \left\{ \begin{pmatrix} 1 & 0 \\ \gamma & 1 \end{pmatrix} : \gamma \in F \right\}$, we can write

$$G = BV \cup Bp. \tag{4.11}$$

Now, if $f \in C_\pi$, then f is completely determined by its restriction to V. Thus, for $v \in V$ and $g = \begin{pmatrix} \alpha & \beta \\ \gamma & \delta \end{pmatrix} \in G$, we may write

$$\begin{aligned} [T^\pi(g)f](v) &= f\left[\begin{pmatrix} 1 & 0 \\ v & 1 \end{pmatrix}\begin{pmatrix} \alpha & \beta \\ \gamma & \delta \end{pmatrix}\right] \\ &= f\left[\begin{pmatrix} \alpha & \beta \\ \alpha v + \gamma & \beta v + \delta \end{pmatrix}\right] \\ &= f\left[\begin{pmatrix} (\beta v + \delta)^{-1} & \beta \\ 0 & \beta v + \delta \end{pmatrix}\begin{bmatrix} 1 & 0 \\ \dfrac{\alpha v + \gamma}{\beta v + \delta} & 1 \end{bmatrix}\right] \end{aligned}$$

where we have identified $v \in V$ with the matrix $\begin{pmatrix} 1 & 0 \\ v & 1 \end{pmatrix}$, $v \in F$, and we have assumed that $\beta v + \delta \neq 0$. Utilizing (4.8) and the expression for Δ_B gives in examples 2.9(iii), we see that

$$[T^{\pi}(g)f](v) = [\text{mod}(\beta v + \delta)]^{-1}\pi(\beta v + \delta)f\begin{bmatrix} 1 & 0 \\ \dfrac{\alpha v + \gamma}{\beta v + \delta} & 1 \end{bmatrix}.$$

Finally, using the fact that V is topologically isomorphic to F, we can identify functions in C_{π} with functions on F by the restriction map. This yields an expression for T^{π} which agrees with the formula (4.3) defining T_{π}. Observe that all the representations T_{π} are infinite-dimensional.

There are many more infinite-dimensional irreducible unitary representations of $SL_2(F)$ in addition to the principal series. However, there is a marked difference between the real and complex cases. For $SL_2(C)$, only the representations in the principal series are needed to derive the Plancherel formula, whereas, for $SL_2(R)$, an infinite collection of irreducible unitary representations, other than those in the principal series, is necessary for the derivation of the Plancherel formula.

The above procedure can also be applied to the group $G = C_2$. Here, we start with non-trivial, one-dimensional unitary representations (that is, characters) of the subgroup $N = \left\{\begin{pmatrix} 1 & z \\ 0 & 1 \end{pmatrix} : z \in C\right\}$. Such characters are in one to one correspondence with the non-zero elements of C, and, for $w \in C^*$, the associated character of N is given by

$$\zeta_w\begin{pmatrix} 1 & z \\ 0 & 1 \end{pmatrix} = e^{iRe(z\bar{w})}, \qquad z \in C. \tag{4.12}$$

Of course, ζ_w may also be regarded as a character on C.

We define C^w to be the space of continuous, complex-valued functions on G satisfying the condition

$$f(ng) = \zeta_w(n)f(g), \quad n \in N, \quad g \in G. \tag{4.13}$$

The representation of G induced by ζ_w acts initially on C^w by the formula

$$[T^w(g')f](g) = f(gg'), \quad g, \quad g' \in G. \tag{4.14}$$

The functions in C^w are completely determined by their restriction to the subgroup $H = \left\{ \begin{pmatrix} 1 & 0 \\ 0 & e^{i\theta} \end{pmatrix} : e^{i\theta} \in T \right\}$. Moreover, any continuous function on H can be extended uniquely to a function in C^w. Since H is topologically isomorphic to T, we can identify C^w with $C(T)$, the space of continuous complex-valued functions on T. Then, for $f \in C^w$, we have

$$f\left[\begin{pmatrix} 1 & 0 \\ 0 & e^{i\varphi} \end{pmatrix}\begin{pmatrix} 1 & z \\ 0 & e^{i\theta} \end{pmatrix}\right] = f\left[\begin{pmatrix} 1 & ze^{-i(\theta+\varphi)} \\ 0 & 1 \end{pmatrix}\begin{pmatrix} 1 & 0 \\ 0 & e^{i(\theta+\varphi)} \end{pmatrix}\right].$$

It follows that, for $g = \begin{pmatrix} 1 & z \\ 0 & e^{i\theta} \end{pmatrix} \in G$ and $f \in C(T)$,

$$[T^w(g)f](e^{i\varphi}) = \zeta_w(ze^{-i(\theta+\varphi)})f(e^{i(\theta+\varphi)}), \qquad e^{i\varphi} \in T. \tag{4.15}$$

Completing $C(T)$ to the Hilbert space $L^2(T)$, we obtain a unitary representation T^w of G on $L^2(T)$ defined by (4.15) for all $f \in L^2(T)$.

PROPOSITION 4.16: *Let $w \in C^*$. Then*

(1) *the map $g \mapsto T^w(g)$ defines an irreducible unitary representation of C_2 on $L^2(T)$;*

(2) *if $|w| = r$, T^w is unitarily equivalent to T^r;*

(3) *the representations $\{T^r;\ r \in R_+^*\}$ together with the representations $\{T_n : n \in Z\}$ given in Examples 3.4 (iii) constitute a complete set of representatives of the unitary dual of C_2.*

It is a simple matter to check (R1), (R2) and (R3). If $w = re^{i\psi}$, then the unitary equivalence stated in (2) is implemented by the map $A_w : L^2(T) \to L^2(T)$ defined by

$$A_w f(e^{i\varphi}) = f(e^{i(\varphi-\psi)}). \tag{4.17}$$

A straightforward computation shows that

$$A_w T^w(g) = T^r(g) A_w \quad \text{for all} \quad g \in G. \tag{4.18}$$

Note that the formula for $T^r(g)$ may be written

$$[T^r(g)f](e^{i\varphi}) = e^{ir[x\cos(\theta+\varphi)+y\sin(\theta+\varphi)]} f(e^{i(\theta+\varphi)}), \tag{4.19}$$

where

$$g = \begin{pmatrix} 1 & z \\ 0 & e^{i\theta} \end{pmatrix}, \quad z = x + iy.$$

The proof of the remaining statements in Proposition 4.16 may be found in the article of Mackey [18] (and the references contained therein). In Mackey's paper, the representation theory of locally compact groups which are semi-direct products is discussed in detail.

We have illustrated the procedure for constructing induced representations above. The importance of this process cannot be overemphasized. For many classes of locally compact groups, all the irreducible unitary representations which are necessary for harmonic analysis on the group can be constructed as induced representations.

5. THE FOURIER TRANSFORM AND CHARACTER THEORY

Let $(T, \mathcal{H})$ be a representation of a locally compact, unimodular group G. For $f \in C_c(G)$, we define the *operator-valued Fourier transform* of f at T by

$$T(f) = \int_G f(x) T(x)\, dx, \tag{5.1}$$

where dx is a Haar measure on G. Thus, $T(f) \in \mathcal{L}(\mathcal{H})$, and, if $(\cdot | \cdot)$ denotes the inner product on $\mathcal{H}$,

$$(T(f)v|w) = \int_G f(x)(T(x)v|w)\, dx \quad \text{for } v,\ \ w \in \mathcal{H}. \tag{5.2}$$

The existence of the above integrals follows from the fact that T is uniformly bounded on compact sets; that is, if K is a compact subset of G, then there exists a constant M such that $\|T(x)\| \leqslant M$ for all $x \in K$.

We are mainly concerned with the case when T is a unitary representation. In this circumstance, $T(f)$ may be defined for all $f \in L^1(G)$. For the remainder of this section, we assume that T is unitary.

For $f, g \in L^1(G)$, the convolution of f and g is defined by

$$(f*g)(y) = \int_G f(x)g(x^{-1}y)\,dx, \quad y \in G. \tag{5.3}$$

Then, $f*g \in L^1(G)$ and $\|f*g\|_1 \leqslant \|f\|_1\|g\|_1$.
There is a natural involution on $L^1(G)$ given by

$$\tilde{f}(x) = \overline{f(x^{-1})}, \quad f \in L^1(G), \quad x \in G. \tag{5.4}$$

It is easy to see that, for $f, g \in L^1(G)$,

$$\begin{aligned} T(f*g) &= T(f)T(g), \\ T(\tilde{f}) &= T(f)^*, \quad \text{the adjoint of } T(f). \end{aligned} \tag{5.5}$$

In order to proceed further, we must consider Hilbert-Schmidt and trace class operators on a separable Hilbert space $\mathcal{H}$. If $\{v_k : k = 0, 1, 2, \dots\}$ is an orthonormal base (ONB) for $\mathcal{H}$ (finite or denumerably infinite) and $A \in \mathcal{L}(\mathcal{H})$, we say that A is a *Hilbert-Schmidt operator* (H-S operator) if $|||A||| = (\Sigma_k\|Av_k\|^2)^{1/2} < \infty$. If A is a H-S operator, then $|||A|||$ is called the Hilbert-Schmidt norm of A, and $|||A|||$ is independent of the choice of the ONB.

If $A \in \mathcal{L}(\mathcal{H})$, then A is of *trace class* if $\Sigma_{j,k}|(Av_k|v_j)| < \infty$. If A is of trace class, then the series $\Sigma_k(Av_k|v_k)$ converges absolutely, and the *trace* of A is

$$\operatorname{tr}(A) = \sum_k (Av_k|v_k). \tag{5.6}$$

This sum is independent of the choice of ONB. For more details about H-S operators and trace class operators, see [5], Ch. XI (particularly pp. 1083–1087) and [24], Part II.

Remark 5.7. Let A be a linear operator on the finite-dimensional Hilbert space C^n, and represent A as an n by n matrix $A = (a_{ij})$ relative to the standard basis in C^n. Then $|||A||| = (\Sigma_{i,j=1}^n |a_{ij}|^2)^{1/2}$, and $\mathrm{tr}(A) = \Sigma_{j=1}^n a_{jj}$.

Now, let $(T, \mathcal{H})$ be an irreducible unitary representation of a locally compact, unimodular group G. As we shall see in the next section, the operator-valued Fourier transform (5.1) and the Hilbert-Schmidt norm on the operators $T(f)$ provide a suitable setting for the Plancherel formula on G. However, in order to obtain a Fourier inversion formula on G, we must have a scalar-valued Fourier transform.

Suppose that f is a complex-valued function on G such that $T(f)$ is of trace class for all $T \in \hat{G}$. Then the map

$$f \mapsto \mathrm{tr}(T(f)) = \hat{f}(T), \quad T \in \hat{G}, \tag{5.8}$$

is called the *scalar-valued Fourier transform* of f. Observe that $\hat{f}(T)$ depends only on the unitary equivalence class of T. For $y \in G$, we set

$$({}^y f)(x) = f(yxy^{-1}), \quad x \in G. \tag{5.9}$$

Then, $T(f)$ is of trace class if and only if $T({}^y f)$ is of trace class for all $y \in G$, and $\hat{f}(T) = ({}^y f)\hat{\ }(T)$ for $T \in \hat{G}$. Thus, the scalar-valued Fourier transform is not injective, in general, since f and ${}^y f$ have the same scalar-valued Fourier transform for all $y \in G$.

If T is a finite-dimensional representation, then

$$\hat{f}(T) = \mathrm{tr}(T(f)) = \int_G f(x)\mathrm{tr}(T(x))\,dx, \tag{5.10}$$

where $\mathrm{tr}(T(x))$ is the trace of an operator on a finite-dimensional space (see Remarks 5.7). We now set

$$\Theta_T(x) = \mathrm{tr}(T(x)), \quad x \in G, \tag{5.11}$$

and call $\Theta_T(x)$ the *character* of T. It is clear that Θ_T is a continuous class function on G. (A function $f: G \to C$ is a *class function* if ${}^y f = f$ for all $y \in G$.) If T is an infinite-dimensional representation, then (5.10) is meaningless since unitary operators on an infinite-dimensional Hilbert space cannot be of trace class. Thus, for infinite-dimensional representations, we cannot take the direct approach which is possible, for example, in the case of compact groups where every irreducible unitary representation is finite-dimensional.

We now restrict our attention to the examples $G = SL_2(F)$, $F = R$ or C, and $G = C_2$, the motion group of the plane. These groups are Lie groups, and, as such, they have an underlying analytic structure. Thus, it makes sense to consider the space $C_c^\infty(G)$ of infinitely differentiable functions in $C_c(G)$. The space $C_c^\infty(G)$ carries a natural topology with respect to which it is a locally convex, complete, Hausdorff, topological vector space ([34], Appendix 2). A continuous linear functional on $C_c^\infty(G)$ is called a *distribution* on G. Note that $f \in C_c^\infty(G)$ if and only if ${}^y f$ is in $C_c^\infty(G)$ for all $y \in G$.

PROPOSITION 5.12: *Let $G = SL_2(F)$ or C_2. Then*
(1) *if $f \in C_c^\infty(G)$, $T(f)$ is of trace class for all $T \in \hat{G}$;*
(2) *for fixed $T \in \hat{G}$, the map $f \mapsto \hat{f}(T)$ is a distribution on G which determines T uniquely up to unitary equivalence.*

The proof of this proposition may be found in [34] where it is proved in a much more general context. For $SL_2(F)$, the reader may also consult [20] and [23].

If Λ is a distribution on G, we say that Λ is an *invariant distribution* if

$$\Lambda(f) = \Lambda({}^y f), \quad f \in C_c^\infty(G), \quad y \in G. \tag{5.13}$$

The most basic example of an invariant distribution is the map $f \mapsto \hat{f}(T)$, $f \in C_c^\infty(G)$, where T is a fixed element of $\hat{G}$. We shall call this the *distribution character* of T. Another important example of an invariant distribution is the *Dirac distribution* δ defined by

$$\delta(f) = f(e), \quad f \in C_c^\infty(G). \tag{5.14}$$

In order to give an explicit form for the distribution characters of the representations constructed in § 4, we shall require the following lemma whose proof may be found in [34], p. 470. This lemma can actually be stated in a much more general measure theoretic context ([5], p. 1086).

LEMMA 5.15: *Let K be a compact Lie group, and let dy be a Haar measure on K normalized so that* $\int_K dy = 1$. *Fix a function* $F \in C^\infty(K \times K)$ *and define a bounded linear operator A on* $L^2(K)$ *by*

$$[Af](x) = \int_K F(x,y)f(y)\,dy, \quad f \in L^2(K), \quad x \in K. \tag{5.16}$$

Then, A is of trace class and $\operatorname{tr}(A) = \int_K F(y,y)\,dy$.

Let us first apply Lemma 5.15 to the representations T^r, $r \in R_+^*$, of C_2 on $L^2(T)$ given by (4.19). If $f \in C_c^\infty(C_2)$, we shall write $f(z, e^{i\theta})$ for $f\begin{pmatrix} 1 & z \\ 0 & e^{i\theta} \end{pmatrix}$. Then, for $\alpha \in L^2(T)$, we have

$$[T^r(f)\alpha](e^{i\varphi})$$

$$= \frac{1}{2\pi}\int_0^{2\pi}\int_C f(z, e^{i\theta})e^{ir[x\cos(\theta+\varphi)+y\sin(\theta+\varphi)]}\alpha(e^{i(\theta+\varphi)})\,dx\,dy\,d\theta$$

$$= \frac{1}{2\pi}\int_0^{2\pi}\alpha(e^{i\theta})\int_C f(z, e^{i(\theta-\varphi)})e^{iRe(zre^{-i\theta})}\,dz\,d\theta$$

$$= \frac{1}{2\pi}\int_0^{2\pi}\alpha(e^{i\theta})F(e^{i\theta}, \; e^{i\varphi})\,d\theta,$$

where

$$F(e^{i\theta},\ e^{i\varphi}) = \int_C f(z, e^{i(\theta-\varphi)}) e^{iRe(zre^{-i\theta})}\, dz$$

is a kernel which satisfies the hypothesis of Lemma 5.15.

Clearly, $F(e^{i\theta}, e^{i\varphi}) = (2\pi)\hat{f}(re^{i\theta}, e^{i(\theta-\varphi)})$, where $\hat{f}$ denotes the usual Fourier transform on the complex plane taken with respect to the first variable. It follows immediately from Lemma 5.15 that

$$\operatorname{tr}(T^r(f)) = \int_0^{2\pi} \hat{f}\,(re^{i\theta}, 1)\, d\theta. \tag{5.17}$$

The computation of the distribution characters of the representations T_π, $\pi \in (F^*)^\wedge$, of $SL_2(F)$ on $L^2(F)$ also utilizes Lemma 5.15. In this case, there is actually a locally integrable function Θ_π on $G = SL_2(F)$ such that

$$\operatorname{tr}(T_\pi(f)) = \int_G f(x)\Theta_\pi(x)\, dx, \qquad f \in C_c^\infty(G). \tag{5.18}$$

Here, dx is a Haar measure on G. The function Θ_π is called the *character* of T_π. A derivation of the characters Θ_π in a general setting may be found in [33]. For $SL_2(F)$, the derivation may be found in [20] and [23]. We shall content ourselves with stating the final results.

From (5.18) and the definition of trace class operators, we see that Θ_π is a class function on G. Consequently, it is only necessary to give a formula for Θ_π on a set of representatives of the conjugacy classes in G. Moreover, as is already indicated by Weyl's character formula for compact Lie groups [34], we should expect an explicit formula for the characters only on a subset of sufficiently "regular" elements in G. For $SL_2(F)$, the set of *regular elements* is the collection of all elements in G which have distinct eigenvalues. The regular elements form a dense, open subset G' of G whose complement has Haar measure zero.

Let A be the subgroup of diagonal elements in G (see (4.5)), and let A' be the set of regular elements in A $(\alpha \neq \pm 1)$. Then, if

$F = C$, every regular element in $SL_2(C)$ is conjugate to an element of A'. On the other hand, if $F = R$, every regular element in $SL_2(R)$ is conjugate either to an element of A' or to an element in the set

$$K' = \left\{ \begin{pmatrix} \cos\theta & -\sin\theta \\ \sin\theta & \cos\theta \end{pmatrix} : \theta \not\equiv 0 (\operatorname{mod} \pi) \right\}. \tag{5.19}$$

Of course, K' is just the set of regular elements in the closed subgroup

$$K = \left\{ \begin{pmatrix} \cos\theta & -\sin\theta \\ \sin\theta & \cos\theta \end{pmatrix} : 0 \leqslant \theta \leqslant 2\pi \right\}.$$

PROPOSITION 5.20: *Let $G = SL_2(C)$ and let $\pi \in (C^*)\hat{}$. Then, the character Θ_π of the representation T_π of G on $L^2(C)$ defined by* (4.3) *is*

$$\Theta_\pi(x) = \frac{\pi(\alpha) + \pi(\alpha^{-1})}{|\alpha - \alpha^{-1}|^2}$$

$$\text{where } x \in G', x \text{ conjugate to } \begin{pmatrix} \alpha & 0 \\ 0 & \alpha^{-1} \end{pmatrix} \in A'.$$

PROPOSITION 5.21: *Let $G = SL_2(R)$ and let $\pi \in (R^*)\hat{}$. Then, the character Θ_π of the representation T_π of G on $L^2(R)$ defined by* (4.3) *is*

$$\Theta_\pi(x) = \begin{cases} \dfrac{\pi(\alpha) + \pi(\alpha^{-1})}{|\alpha - \alpha^{-1}|}, & \text{where } x \in G', \\ & x \text{ conjugate to } \begin{pmatrix} \alpha & 0 \\ 0 & \alpha^{-1} \end{pmatrix} \in A', \\ 0, & \text{if } x \text{ is conjugate to an element of } K'. \end{cases}$$

Thus, we see that the characters of the representations in the principal series of $SL_2(C)$ and $SL_2(R)$ have a similar form on A'.

However, as we pointed out in § 4, there is a marked difference between $SL_2(C)$ and $SL_2(R)$. For $SL_2(C)$, the conjugates of A' fill out all of the group except for a set of measure zero. For this reason, the representations of the principal series and the characters of these representations are sufficient for most aspects of harmonic analysis. On the other hand, for $SL_2(R)$, there is no chance that the representations of the principal series could be sufficient for harmonic analysis since the characters of these representations have no support on the open set consisting of conjugates of K'. Consequently, irreducible unitary representations other than those in the principal series are required.

Finally, it follows from Propositions 5.12, 5.20 and 5.21 that the representations T_π and $T_{\pi^{-1}}$ in the principal series are unitarily equivalent. It is possible to give an explicit intertwining operator for these two representations, but this operator plays no role in our development. We refer the reader to [14], [15] and [13] for the construction of these operators and more detailed information about the role of intertwining operators in representation theory.

6. PLANCHEREL'S FORMULA AND FOURIER INVERSION

In order to motivate the discussion in this section, we first recall the classical formulation of Plancherel's theorem and Fourier inversion for R and T (for T, Plancherel's theorem is usually referred to as Parseval's identity).

PLANCHEREL'S THEOREM [21]: *Suppose that $f \in L^2(R)$ and set*

$$S_N(t) = \left(1/\sqrt{2\pi}\right)\int_{-N}^{N} f(x)e^{itx}\,dx,$$

where N is a positive integer. Then, $S_N \in L^2(R)$, and the sequence $(S_N)_{N=1}^{\infty}$ is a Cauchy sequence in $L^2(R)$. Thus, there is a function $\hat{f} \in L^2(R)$ such that

$$\lim_{N\to\infty}\int_{-\infty}^{\infty} |\hat{f}(t) - S_N(t)|^2\,dt = 0.$$

Moreover, if we set

$$\hat{S}_N(x) = (1/\sqrt{2\pi})\int_{-N}^{N} \hat{f}(t)e^{-itx}\,dt,$$

then

$$\lim_{N\to\infty}\int_{-\infty}^{\infty} |f(x) - \hat{S}_N(x)|^2\,dx = 0.$$

Finally, the Fourier transform is a unitary map on $L^2(R)$; that is, $\|f\|_2 = \|\hat{f}\|_2$.

PARSEVAL'S IDENTITY [37]: *Suppose that $f \in L^2(T)$ and set*

$$\hat{f}(n) = (1/2\pi)\int_0^{2\pi} f(e^{i\theta})e^{-in\theta}\,d\theta, \quad n \in Z.$$

Then,

$$\sum_{n=-\infty}^{\infty} |\hat{f}(n)|^2 < \infty,$$

and

$$\sum_{n=-\infty}^{\infty} |\hat{f}(n)|^2 = (1/2\pi)\int_0^{2\pi} |f(e^{i\theta})|^2\,d\theta.$$

Moreover, if $(c_n)_{n=-\infty}^{\infty}$ is a sequence of complex numbers such that $\sum_{n=-\infty}^{\infty}|c_n|^2 < \infty$, then there is a unique element $f \in L^2(T)$ such that $\hat{f}(n) = c_n$ for all n.

In both cases above, there is a positive measure on the unitary dual $\hat{G}$ of the group G (either R or T) such that the Fourier transform on G is a unitary mapping from $L^2(G)$ to $L^2(\hat{G})$. In addition, a function $f \in L^2(G)$ can be recovered (in the L^2 sense) from its Fourier transform.

In the classical theory, Fourier inversion signifies the recovery of a function in the pointwise sense from its Fourier transform. For this type of result, it is necessary to require some smoothness properties. Thus, if $f \in C_c^\infty(R)$, then $\hat{f} \in C^\infty(R)$ and vanishes rapidly at infinity, and

$$f(x) = \left(1/\sqrt{2\pi}\right)\int_{-\infty}^{\infty} \hat{f}(t)e^{-itx}\,dt, \quad x \in R. \tag{6.1}$$

Also, if $f \in C^\infty(T)$, then

$$f(e^{i\theta}) = \sum_{n=-\infty}^{\infty} \hat{f}(n)e^{in\theta}, \quad e^{i\theta} \in T, \tag{6.2}$$

where the series converges absolutely and uniformly to f on T. The fact that the Fourier transform is a unitary mapping from $L^2(G)$ to $L^2(\hat{G})$ can be recovered from the Fourier inversion formula in a manner to be illustrated below.

In order to obtain a suitable generalization of Plancherel's theorem and Parseval's identity to a locally compact group G, we must have both a measure on $\hat{G}$ and a norm on the operators $T(f)$, $T \in \hat{G}$, which is analogous to the L^2 norm. The required norm is precisely the Hilbert-Schmidt norm introduced in § 5. At this point, we must make some assumptions about the group G in order to state the Plancherel theorem. Thus, we assume that G is a locally compact, separable, unimodular group which is "type I" (or *postliminaire*). This latter condition is somewhat technical, and we shall not elaborate further upon it (for a complete discussion, see [4]). Suffice it to say that our examples, $SL_2(F)$ and C_2 satisfy all of the above conditions.

THEOREM 6.3 (Plancherel's theorem): *Let G be a locally compact group which satisfies the above conditions, and let dx be a fixed Haar measure on G. Then, there exists a positive measure μ on $\hat{G}$ (determined uniquely up to a constant which depends only on the normalization of dx) such that, for $f \in L^1(G) \cap L^2(G)$, $T(f)$ is a* H-S *operator for μ-almost all $T \in \hat{G}$, and*

$$\int_G |f(x)|^2\,dx = \int_{\hat{G}} |||T(f)|||^2 d\mu(T). \tag{6.4}$$

The measure μ is called the *Plancherel measure* for G. The general Plancherel theorem goes back to Segal [28]. The proof of Theorem 6.3 may be found in [4], p. 328.

Note that Theorem 6.3 yields no information about the specific form of the Plancherel measure. One of the major problems in representation theory over the past two decades has been the explicit determination of the Plancherel measure for particular classes of groups.

Examples 6.5. (i) If we take the Haar measure $dx/\sqrt{2\pi}$ on $G = R$, then it follows from the classical Plancherel's theorem that the Plancherel measure on $\hat{G} = R$ is given by $dt/\sqrt{2\pi}$. Also, if we take the Haar measure $d\theta/2\pi$ on $G = T$, then it follows from Parseval's identity that the Plancherel measure on $\hat{G} = Z$ is given by the measure which assigns a mass equal to one to each point in Z.

(ii) If G is a compact group, then the Plancherel formula for G is $\|f\|^2 = \sum_{T\in\hat{G}} d_T |||T(f)|||^2$, $f \in L^2(G)$ (see the article of Weiss). Thus, the Plancherel measure on $\hat{G}$ assigns the mass d_T = degree of T to each irreducible, unitary representation $T \in \hat{G}$. (Recall that $d_T < \infty$.)

(iii) Let $G = SL_2(C)$. If $\pi \in (C^*)\hat{}$, then, using (4.2), we write $T_\pi = T_{n,s}$, $n \in Z$, $s \in R$. The support of the Plancherel measure for G is the set $\{T_{n,s} : n = 0 \text{ and } s \geqslant 0, \text{ or } n > 0 \text{ and } s \in R\}$ (recall that $T_{n,s}$ is unitarily equivalent to $T_{-n,-s}$). However, in stating the Plancherel formula for G, we find it simpler for indexing purposes to take $\{T_{n,s} : n \in Z,\ s \in R\}$ so that (up to unitary equivalence) each representation in the support of the Plancherel measure occurs twice. Then, for a suitably normalized Haar measure dx on G and $f \in L^1(G) \cap L^2(G)$,

$$\int_G |f(x)|^2\, dx = \frac{1}{32\pi^4} \sum_{n=-\infty}^{\infty} \int_{-\infty}^{\infty} |||T_{n,s}(f)|||^2 (n^2 + s^2)\, ds, \quad (6.6)$$

where ds is Lebesgue measure on R. Thus, the Plancherel measure for G is $(n^2 + s^2)\, ds$. A proof of (6.6) may be found in [20], §14.

We now turn to the problem of Fourier inversion. Assume that $G = SL_2(F)$ or C_2, and let Λ be an invariant distribution on G (see (5.13)). The *Fourier transform* of Λ is a linear functional $\hat{\Lambda}$ on $\hat{G}$ such that

$$\Lambda(f) = \hat{\Lambda}(\hat{f}), \quad f \in C_c^\infty(G). \tag{6.7}$$

Fourier inversion on G takes the following form:

Given an invariant distribution Λ on G, find $\hat{\Lambda}$.

For example, if $\Lambda = \delta$, the Dirac distribution (5.14), then $\hat{\Lambda}$ is just the Plancherel measure for G, and we have

$$\delta(f) = f(e) = \int_{\hat{G}} \hat{f}(T)\,d\mu(T), \quad f \in C_c^\infty(G), \tag{6.8}$$

where μ is the Plancherel measure on $\hat{G}$. The formula (6.8) is often called the *Plancherel formula* for G. For a further discussion of the Fourier transform of invariant distributions and some additional examples, see [27].

Using the Plancherel formula (6.8), we can derive (6.4) in a simple fashion. If $f \in C_c^\infty(G)$, then $\tilde{f}$ and $f * \tilde{f}$, the convolution of f and $\tilde{f}$, are in $C_c^\infty(G)$. Moreover

$$(f*\tilde{f})(e) = \int_G f(x)\tilde{f}(x^{-1})\,dx$$

$$= \int_G |f(x)|^2\,dx.$$

On the other hand, it follows from (5.5), (5.6) and (5.10) that

$$(f*\tilde{f})\hat{\ }(T) = \operatorname{tr}(T(f*\tilde{f})) = \operatorname{tr}(T(f)T(\tilde{f}))$$

$$= \operatorname{tr}(T(f)T(f)^*) = |||T(f)|||^2.$$

Thus, from (6.8), we obtain

$$\int_G |f(x)|^2\,dx = (f*\tilde{f})(e)$$

$$= \int_{\hat{G}} (f*\tilde{f})\hat{\ }(T)\,d\mu(T)$$

$$= \int_{\hat{G}} |||T(f)|||^2 d\mu(T).$$

Let us now derive the Plancherel formula for C_2. From (5.17), we have $\mathrm{tr}(T^r(f)) = \int_0^{2\pi} \hat{f}(re^{i\theta}, 1)\,d\theta$, where

$$\hat{f}(re^{i\theta}, 1) = \frac{1}{2\pi}\int_C f(z, 1)e^{iRe(zre^{-i\theta})}\,dz.$$

Then, by the standard Fourier inversion formula for C, we have (using integration in polar coordinates)

$$f(0, 1) = \frac{1}{2\pi}\int_C \hat{f}(w, 1)\,dw$$

$$= \frac{1}{2\pi}\int_0^{2\pi}\int_0^{\infty} \hat{f}(\rho e^{i\zeta}, 1)\rho\,d\rho\,d\zeta.$$

Thus,

$$f(e) = f(0, 1) = \frac{1}{2\pi}\int_0^{\infty} \mathrm{tr}(T^r(f))r\,dr. \tag{6.9}$$

This is the Plancherel formula for C_2. The Plancherel measure is $r\,dr$.

7. A BRIEF INTRODUCTION TO THE LITERATURE

The origins of harmonic analysis on locally compact groups are rooted in the eighteenth century. However, the study of infinite-dimensional representations of locally compact groups began only about thirty-five years ago, and substantial developments took place soon after the end of World War II. Roughly speaking, these developments followed two separate paths which have continued up to the present day. The first might be called "general representation theory" in which as few restrictions as possible are placed on the locally compact groups under consideration. In this context, one gets "general" versions of Plancherel's theorem (Theorem 6.3), character theory, induced representations, topology on the unitary dual and much more. A good introduction to this aspect of representation theory may be found in Dixmier's book [4] (which also contains an extensive bibliography), and in Mackey's notes [17]. The notes of Auslander [1] might also prove helpful in this direction.

In the development of representation theory, the second path, which has been predominant over the past fifteen years, is the study of the representation theory of specific classes of groups such as semi-direct products, nilpotent Lie groups, solvable Lie groups, and semisimple (and reductive) Lie groups. More recently, groups defined over p-adic fields have received considerable attention. Here, the structure of the groups enters into Plancherel's formula, character theory, etc., in an essential way, and the nature of the subject is quite different from "general representation theory". Much of this work is summarized, mostly without proofs, in the Proceedings of the Symposia in Pure Mathematics, v. XXVI, Harmonic analysis on homogeneous spaces, edited by C. C. Moore, AMS, 1973.

It is often useful to study the representation theory of one particular group or a small subclass of groups within one of the classes listed above in order to obtain better understanding and motivation. For example, in the case of semisimple Lie groups, the representation theory of $SL_2(R)$ and $SL_2(C)$ provides a good starting point (for $SL_2(R)$, see [2], [6], [7], [11], [14], [23], [30]; for $SL_2(C)$, see [7], [20]). Also, for the classical, complex, semisimple Lie groups, there is the book of Gel'fand and Neumark [8].*

---

*Added in proof: The recent book, $SL_2(R)$, by S. Lang is a very good introductory text.

For the general theory of semi-direct products, the article of Mackey in [18] is a good source. Pukanszky's book [24] contains a treatment of nilpotent Lie groups along with a brief introduction to some of the elementary aspects of representation theory. The general theory of representations of semisimple Lie groups, due almost exclusively to Harish-Chandra, is covered in the massive work of Warner [34] (there have been many further developments since the publication of Warner's work). Since Warner's tomes are rather forbidding (in terms of both prerequisites and technicality), the enthusiastic beginner might wish to start by consulting the works of Borel [3], Gross [10], Lipsman [16], Varadarajan [31], Vilenkin [32] and Wallach [33].

The literature in the field of representation theory has expanded enormously in the past decade. A good idea of the growth of the subject can be gleaned from the bibliography in Warner [34]. All in all, it is a difficult task to get into the area. The author's lecture notes [26] might prove helpful in obtaining an overview.

Finally, in order to maintain some perspective in this avalanche, the reader should go back to the articles of Stone [29] and Wiener [36], and the book of Zygmund [37] for the reflections of these masters on the foundations of harmonic analysis.

REFERENCES

1. Auslander, L., *Unitary Representations of Locally Compact Groups—The Elementary and Type I Theory*, Lecture Notes, Yale University, 1962.

2. Bargmann, V., "Irreducible unitary representations of the Lorentz group," *Ann. of Math.*, **48** (1947), 568–640.

3. Borel, A., *Représentations de Groupes Localement Compacts*, Springer Lecture Notes in Mathematics, v. 276, 1972.

4. Dixmier, J., *Les C*-algèbres et leurs représentations*, Gauthier-Villars, Paris, 1964.

5. Dunford, N., and J. T. Schwartz, *Linear Operators*, Part II, Interscience, New York, 1963.

6. Gel'fand, I. M., M. I. Graev and I. I. Pyateckii-Shapiro, *Representation theory and automorphic functions*, Saunders, Philadelphia, 1969.

7. Gel'fand, I. M., M. I. Graev and N. Ya. Vilenkin, *Generalized Functions*, v. 5, Academic Press, New York, 1966.

8. Gel'fand, I. M., and M. A. Neumark, *Unitäre Darstellungen der Klassischen Gruppen*, Akademie Verlag, Berlin, 1957.

9. Godement, R., "A theory of spherical functions, I," *Trans. Amer. Math. Soc.*, **73** (1952), 496–556.

10. Gross, K., *Representations of Locally Compact Groups*, to be published.

11. Harish-Chandra, "Plancherel formula for the 2×2 real unimodular group," *Proc. Nat. Acad. Sci. USA*, **38** (1952), 337–342.

12. Hewitt, E., and K. A. Ross, *Abstract Harmonic Analysis*, v. I, Springer, Berlin, 1963.

13. Knapp, A. W., and E. M. Stein, "Intertwining operators for semi-simple groups," *Ann. of Math.*, **93** (1971), 489–578.

14. Kunze, R. A., and E. M. Stein, "Uniformly bounded representations and harmonic analysis on the 2×2 real unimodular group," *Amer. J. Math.*, **82** (1960), 1–62.

15. Kunze, R. A., and E. M. Stein, "Uniformly bounded representations II: Analytic continuation of the principal series of the $n \times n$ complex unimodular group," *Amer. J. Math.*, **83** (1961), 723–786.

16. Lipsman, R., *Group Representations*, Springer Lecture Notes in Mathematics, v. 388, 1974.

17. Mackey, G., *The Theory of Group Representations*, Lecture Notes, University of Chicago, 1955. (Republished by University of Chicago Press, 1976.)

18. Segal, I. E., *Mathematical Problems of Relativistic Physics*, Amer. Math. Soc., Providence, R. I., 1963 (Appendix by G. Mackey entitled *Group Representations in Hilbert Space*).

19. Nachbin, L., *The Haar Integral*, Van Nostrand, New York, 1965.

20. Naimark, M. A., *Linear Representations of the Lorentz Group*, Macmillan, New York, 1964.

21. Plancherel, M., "Contribution à l'étude de la représentation d'une fonction arbitraire par les intégrales définies," *Rend. Circ. Mat. Palermo*, **30** (1910), 289–335.

22. Pontryagin, L. S., *Topological Groups*, 2nd edition, Gordon and Breach, New York, 1966.

23. Pukanszky, L., "The Plancherel formula for the universal covering group of $SL(R, 2)$," *Math. Ann.*, **156** (1964), 96–143.

24. Pukanszky, L., *Leçons sur les représentations des groupes*, Dunod, Paris, 1967.

25. Rudin, W., *Fourier Analysis on Groups*, Interscience, New York, 1962.

26. Sally, P., *Harmonic analysis on locally compact groups*, Lecture Notes, University of Maryland, 1976.

27. Sally, P., and G. Warner, *The Fourier transform of invariant distributions, in Conference on Harmonic Analysis*, Springer Lecture Notes in Mathematics, v. 266, 1972.

28. Segal, I. E., "An extension of Plancherel's formula to separable unimodular groups," *Ann. of Math.*, **52** (1950), 272–292.

29. Stone, M., "On the foundations of harmonic analysis," *Medd. Lunds. Univ. Mat. Sem., Tome suppl.*, (1952), 207–227.

30. Takahashi, R., "Sur les fonctions sphériques et la formule de Plancherel dans les groupes hyperboliques," *Japan. J. Math.*, **31** (1961), 55–90.

31. Varadarajan, V. S., *Lie Groups, Lie Algebras and their Representations*, Prentice-Hall, Englewood Cliffs, N. J., 1974.

32. Vilenkin, N. Ya., *Special Functions and the Theory of Group Representations*, Amer. Math. Soc., Providence, R. I., 1968.

33. Wallach, N., *Harmonic Analysis on Homogeneous Spaces*, Dekker, New York, 1973.

34. Warner, G., *Harmonic Analysis on Semi-Simple Lie Groups*, 2 volumes, Springer, Berlin, 1972.

35. Weil, A., *L'intégration dans les groupes topologiques et ses applications*, Hermann, Paris, 1940.

36. Wiener, N., *The historical background of harmonic analysis*, AMS Semicentennial Publications, v. II, Amer. Math. Soc., New York, 1938, pp. 56–68.

37. Zygmund, A., *Trigonometric Series*, v. I, Cambridge University Press, Cambridge, 1959.

HARMONIC ANALYSIS ON CARTAN AND SIEGEL DOMAINS

Stephen Vági

1. INTRODUCTION

This lecture surveys, informally, and rather sketchily, the development of those aspects of function theory on Cartan and Siegel domains which are of interest to the harmonic analyst, or to be more precise, to the *classical* harmonic analyst: Poisson and Cauchy integrals, the conjugate function operator, Hardy spaces, and the boundary behavior of holomorphic and harmonic functions. In the one variable theory the geometric properties of the unit disc and the upper half-plane play a subordinate role, or rather the relevant geometry is so simple that it requires no separate study. This is no longer true in the case of several variables; nor even, let me hasten to add, in the case of one variable if one's interest is not in Fourier series or H^p spaces, but in automorphic functions or Riemann surfaces. Therefore in dealing with several variables it becomes necessary to heed the advice of Hermann Weyl, and to take a closer look at the "soil in which the functions grow" before discussing function theory proper. Accordingly a large portion of this exposition is devoted to the domains as such.

2. THE DOMAINS

Omitting from its statement any reference to the unit disc, one can state the Riemann mapping theorem in the following form: *Any two simply connected proper subdomains (a domain is a connected open set) of the complex plane are holomorphically equivalent.* Since the only other simply connected domain in the plane is the plane itself, this theorem can be regarded as giving a classification of the simply connected plane domains relative to holomorphic equivalence: there are two equivalence classes—one consisting of one element only (viz., the plane), the other one comprising all the other domains, its standard representative being the unit disc. It can be said that our subject was born when Poincaré discovered in 1907 that the Riemann mapping theorem fails to extend to higher dimensions. He proved that the Cartesian product of two discs is not holomorphically equivalent to the unit ball in C^2 [68].

Quite recently H. Alexander found a very short elementary proof of this result [1], and since it uses the theory of Hardy spaces I shall include it. Set $U^2 = \{z \in C^2 | \, |z_1| < 1, |z_2| < 1\}$ and $B^2 = \{z \in C^2 | \, |z_1|^2 + |z_2|^2 < 1\}$. If $F = (f_1, f_2) : U^2 \to B^2$ is a holomorphic map, set

$$\|F(z)\|^2 = |f_1(z)|^2 + |f_2(z)|^2 \quad \text{and} \quad F^1(z) = F(z_1, 0).$$

Since $F(z) \in B_2$, f_1 and f_2 are bounded and *a fortiori* belong to $H^2(U^2)$. In particular they have boundary values on $T^2 = \{z \in C^2 | \, |z_1| = |z_2| = 1\}$ which is the Šilov (or distinguished) boundary of U^2. Poincaré's theorem easily follows from the following inequality which we temporarily take for granted:

$$\int_0^{2\pi} \int_0^{2\pi} \|F(z) - F^1(z)\|^2 \, d\theta_1 \, d\theta_2 \leqslant 2\pi \int_0^{2} \left(1 - \|F^1(z)\|^2\right) d\theta_1,$$

where $z = (r_1 e^{i\theta_1}, r_2 e^{i\theta_2})$ with $0 \leqslant r_i < 1$ for $i = 1, 2$. In fact, suppose that F is a homeomorphism of U^2 onto B^2. Then for $z \in T^2$, $|F^1(rz)| \to 1$ as $r \to 1$, because $(rz_1, 0)$ tends to the boundary of U^2. Therefore the right side of the inequality is zero, and consequently $F = F^1$. This means that F depends only on the variable z_1, in particular it is not one to one. This contradicts the fact that F is a homeomorphism. To establish the inequality, note that for any g belonging to $H^2(U^2)$ $z_2 \mapsto g(z_1, z_2)$ for fixed z_1 is holomorphic, and therefore has the mean value property. Hence

$$\frac{1}{2\pi} \int_0^{2\pi} g(z_1, z_2)\, d\theta_2 = g(z_1, 0).$$

Multiply this by $\overline{g(z_1, 0)}$ and integrate to obtain

$$\int_0^{2\pi} \int_0^{2\pi} g(z)\, \overline{g(z_1, 0)}\, d\theta_1\, d\theta_2 = \int_0^{2\pi} \int_0^{2\pi} |g(z_1, 0)|^2\, d\theta_1\, d\theta_2.$$

Apply this to obtain

$$\int_0^{2\pi} \int_0^{2\pi} |f_i(z) - f_i(z_1, 0)|^2\, d\theta_1\, d\theta_2$$

$$= \int_0^{2\pi} \int_0^{2\pi} |f_i(z)|^2\, d\theta_1\, d\theta_2 - \int_0^{2\pi} \int_0^{2\pi} |f_i(z_1, 0)|^2\, d\theta_1\, d\theta_2, \quad i = 1, 2.$$

Add these equalities for $i = 1, 2$ to obtain

$$\int_0^{2\pi} \int_0^{2\pi} \|F(z) - F^1(z)\|^2\, d\theta_1\, d\theta_2$$

$$= \int_0^{2\pi} \int_0^{2\pi} \left(\|F(z)\|^2 - \|F^1(z)\|^2\right) d\theta_1\, d\theta_2,$$

and recall that $\|F(z)\|^2 \leqslant 1$.

Poincaré's discovery raises the question of classifying up to holomorphic equivalence the simple connected domains in complex n-space. This is a problem of very great difficulty. Function theorists would be happy with much less; moreover they are primarily interested in *domains of holomorphy*, i.e., domains on which there are holomorphic functions which cannot be continued analytically into any larger domain. From the point of view of function theory these are the "natural" domains. In the plane all domains, simply connected or not, are domains of holomorphy. In higher dimensions—and this is another distinctive feature of the several variable theory—this is no longer so. There are even simply connected domains which are not domains of holomorphy, a spherical shell in C^n is an example. (Both the polydisc and the ball, however are domains of holomorphy.) Even the more limited program of classifying the simply connected domains of holomorphy turns out to be very hard. Further restrictions have to be imposed, restrictions which will effectively simplify the problem without limiting the scope of the resulting theory too much. The theory should be rich enough to apply to interesting situations.

Now, in addition to being simply connected and a domain of holomorphy, the disc has other, deeper properties which are equally important in function theory. The most important of these is *homogeneity*: given any two points z and w of the disc U, there exists a holomorphic automorphism (i.e., a holomorphic homeomorphism of U onto itself) which carries z into w. Every automorphism is of the form $z \mapsto e^{i\alpha}(z - a)/(1 - \bar{a}z)$ where $a \in U$ and α is real. The automorphisms form a group which is essentially isomorphic to the group of 2×2 matrices of determinant one. To say that the disc is homogeneous means the same thing as saying that the group of automorphisms acts *transitively* on it. Next in importance is the fact that, in addition to its usual (Euclidean) metric which it inherits from the surrounding plane, the disc is endowed with another, non-Euclidean metric, the hyperbolic or Poincaré metric, which is, in fact, defined in function theoretic terms. Moreover the holomorphic automorphisms of U are the isometries of the Poincaré metric. Finally there is a third property, *symmetry*, which is of a more special nature; its discussion will be taken up a little later. Homogeneity and the Poincaré metric are of paramount importance for the role which the disc plays in the theory of automorphic functions and in the theory of

Riemann surfaces. It is the universal covering surface of some of the most important Riemann surfaces, in particular of certain compact surfaces (those of genus bigger than one[1]). Let me note further that one of the interesting problems in passing from one to several variables is to find out what the analogues of compact Riemann surfaces are and to what extent a theory similar to the one variable theory can be constructed. All these circumstances reasonably justify the claim that *bounded homogeneous domains* have on the interest of function theorists; in short, they constitute a class of domains sufficiently wide and interesting to study in depth. The fact that their function theory offers a fertile field of application for harmonic analysis constitutes their interest in the present context. Let me add that these domains are simply connected, they are domains of holomorphy and they all have a natural metric linked to function theory, the Bergman metric, which for the disc coincides with the hyperbolic metric. Considerable progress has been made in their study, but a complete structure theory has not yet been established.

Henri Cartan [10] undertook a systematic study of these domains and obtained the first deep results about them. The most important one is that the group of automorphisms of such a domain is a Lie group.[2] This discovery brought the subject within the scope of differential geometry and attracted Élie Cartan to it. Before describing his contributions we must take a look at the third property of the circle I mentioned before (but did not discuss): symmetry. Let σ denote the automorphism of the disc U which carries z into $-z$; σ has two quite obvious properties: it leaves the origin, and no other point fixed, and its square is the identity. Let now a be any point of U, g the automorphism which carries a to 0. Then $\sigma_a = g^{-1}\sigma g$ is an automorphism which leaves

[1]I should mention here that in the second half of the 19th century problems in the theory of compact Riemann surfaces provided the strongest stimulus for the study of functions of several complex variables.

[2]The condition that the domain be bounded cannot be dropped. W. Kaup [36] has given examples of homogeneous unbounded domains on which no Lie group acts transitively.

a, and no other point fixed, and its square is the identity. Essentially, the state of affairs described in the last sentence constitutes symmetry. More formally: a domain $D \subset C^n$ is *symmetric* if for every $a \in D$ there is a holomorphic automorphism σ_a of D of which a is an isolated fixed point, and whose square is the identity. (A non-trivial selfmapping of a set whose square is the identity is called an involution.) It is a remarkable fact that this property which at first sight appears rather unassuming has farreaching consequences. Here are a few: a bounded symmetric domain is simply connected, homogeneous, and its group of automorphisms is a semisimple Lie group. These domains are also called Cartan domains, after Élie Cartan who introduced them in 1935 in the fundamental paper *Sur les domaines bornés homogènes de l'espace de n variables complexes* [9]. Recall that a bounded domain has a natural Riemannian metric, the Bergman metric. Therefore the additional condition of symmetry turns the domain into a Riemannian symmetric space. This type of space had been introduced in the late twenties by É. Cartan who then studied them intensively. A considerable amount of information about them was available; Cartan now brought to bear this accumulated knowledge on the study of symmetric domains. He found all of them and classified them.

The result of the classification can be described in fairly simple terms. If D_1 and D_2 are symmetric domains, then so is their cartesian product $D_1 \times D_2$. If the domain is not the cartesian product of lower dimensional domains it is said to be *irreducible*. Every domain is uniquely (up to the order of its factors) the product of irreducible ones. There are four classes of irreducible domains, each containing infinitely many members, and two isolated domains of dimensions 16 and 27, respectively. The domains of the four classes are the so-called *classical domains*. The two others are called the *exceptional domains*. The four classes turn out to be composed of domains whose standard realizations are quite familiar objects; three of them being composed of matrix domains.

Let $M_{m,n}$ denote the set of $m \times n$ complex matrices and set $M_{n,n} = M_n$. If $Z \in M_{m,n}$ let Z^* denote the conjugate transpose matrix of Z, and let I always denote the identity matrix of appropriate size. If $A \in M_n$, $A > 0$ signifies that A is a positive definite hermitian matrix. Every irreducible classical domain is holomorphically equivalent to one of the domains listed below; these are the standard representatives of their classes:[3]

Class I: $\{Z \in M_{m,n} | I - Z^*Z > 0\}$, $m \geqslant n \geqslant 1$.

Class II: $\{Z \in M_n | Z$ is symmetric, and $I - Z^*Z > 0\}$, $n \geqslant 2$.

Class III: $\{Z \in M_n | Z$ is skew symmetric, and $I - Z^*Z > 0\}$, $n \geqslant 5$.

Class IV: $$z \in C^n \Big| \left|\sum_{k=1}^{n} z_k^2\right|^2 - 2\sum_{k=1}^{n} |z_k|^2 + 1 > 0,$$

$$\text{and } 1 - \left|\sum_{k=1}^{n} z_k^2\right| > 0 \Big\}, \quad n \geqslant 5.$$

The restrictions on the dimensions aim at avoiding overlaps between different classes, e.g., for $m = n = 1$ the domain of class I is clearly the unit disc, but so would be the domain of class IV if $n = 1$ were allowed, etc. Also for low dimensions some of the domains listed would be reducible, e.g., this would occur for the domain of class IV if $n = 2$ were permitted. The polydisc U^k, $k > 1$, is the product of k domains of class I with $n = m = 1$, it is a reducible domain. The unit ball in C^k is irreducible (this gives a new proof of the fact that it is inequivalent to U^k), it is the domain of class I with $m = k$, $n = 1$. The domains of class IV, the Lie spheres, look less familiar, but it can be shown that they have another (unbounded) realization, which is very simple—viz., $\{z = x + iy \in C^n | y_n > +\sqrt{\Sigma_{k=1}^{n-1} y_k^2}\}$, the "tube domain" over the cone $\{y \in R^n | y_n > +\sqrt{\Sigma_{k=1}^{n-1} y_k^2}\}$, (We shall see more of these

[3]The numbering of the classes follows that used in the book of Hua [28] and is standard in the subject. In Cartan's own classification the order is I, III, IV, II.

later.) All the domains, including the exceptional ones can be realized as matrix domains, as has been shown by M. Ise [31], [32], [33]. For the purposes of this exposition the exceptional domains will be consistently ignored. There exists a special literature about them. (See, e.g., [3] where more sources are quoted.) The presence of these domains on Cartan's list is rather irksome, and is one of the reasons why the approach based on the general theory of (Riemannian) symmetric spaces and Lie groups is indispensable to prove theorems valid for *all* domains. Quite frequently results for the classical domains can be established by separate case by case consideration of the four classes, and in general, if a result has been proved for one class of domains, the proofs for the other classes can be expected to follow somewhat similar patterns. The early workers in the subject relied on this method and usually ignored the exceptional domains whose dimensions are high enough to make explicit calculations difficult and very tiresome. This fact, of course, also accounts for the existence of a special literature devoted to them.

In his 1935 paper Cartan also proved that for dimensions two and three, all homogeneous domains are symmetric.[4] (For dimension two there are only two domains: U^2 and B^2—the unit ball; for $n = 3$ the list is: U^3, B^3, $U \times B^2$, and the Lie sphere.) Cartan then raised the question whether this was the case in higher dimensions, too. The efforts to answer this question constitute an important strand in the history of the subject up to 1959 when I. I. Pjateckiĭ-Šapiro showed that the answer is negative. The bounded symmetric domains are the "proper" generalizations of the unit disc in higher dimensions, their structure is quite well understood, and they have manifold connections with other fields of mathematics. All this makes them into a very important subclass of the larger class of bounded homogeneous domains.

Let me now return to the subsequent history of Cartan's problem, and of the classification problem in general. In 1954 Borel [6] and Koszul [55] showed that a bounded domain which has a

---

[4]His paper contains the proof only for $n = 2$. He did not include the proof for $n = 3$, because the calculations were too extensive.

transitive semisimple automorphism group is a bounded symmetric domain. Hano in 1957 [25] strengthened this result by showing that unimodularity can be substituted for semisimplicity in the preceding statement. Finally in 1959 Pjateckiĭ-Šapiro [65], [66], constructed the first example of a non-symmetric bounded homogeneous domain in C^4. It turns out that symmetric domains are, really, rather exceptional [67]. For $n \geqslant 7$ there are continuum many holomorphically inequivalent homogeneous domains in C^n, but only finitely many inequivalent symmetric ones.[5]

Cartan after having stated his problem commented that the approach he used in dimensions two and three was unlikely to work in higher dimensions, and that one would have to "*s'appuyer sur une idée nouvelle*". The new idea which led to the solution was to investigate Siegel domains (the term is due to Pjateckiĭ-Šapiro). These domains, or at any rate one of them, made their first appearance in Siegel's famous paper *Symplectic*[6] *Geometry* [75] which he wrote to prepare the ground for the investigation of automorphic functions of several variables. It is worth pointing out that Pjateckiĭ-Šapiro's interest in these domains was also motivated by the theory of automorphic functions. This is also true of L. K. Hua, another pioneer of the subject whose paper [29] appeared almost simultaneously with Siegel's.

Just as bounded symmetric domains are generalizations of the unit disc, so Siegel domains are generalizations of the upper half-plane and therefore are also frequently called (generalized) half-planes. The unit disc is holomorphically equivalent to the upper half-plane; a mapping is given by $z \mapsto i(1 - z)/(1 + z)$. (It is not unique.) Every Cartan domain has a realization as a Siegel domain. The corresponding mapping is called the *Cayley transform*. However, it is important to note that the class of Siegel

[5]Though finite their number rapidly gets quite large, e.g., for $n = 12$ there are 300.

[6]The automorphism group of the domain considered by Siegel is the so-called *symplectic group*, which accounts for the title of his paper. The name "symplectic group" for the group in question had been proposed by Hermann Weyl in his book on the classical groups [88, p. 163].

domains is not exhausted by the half-plane realizations of Cartan domains. The domain originally considered by Siegel is one equivalent to a bounded domain of class II. For the classical domains the half-plane realizations and Cayley transforms are obtained in Pjateckiĭ-Šapiro by case by case consideration of the four classes. In some cases the Cayley transform was known earlier—for class II it is in Siegel's paper [75], for class IV it appears explicitly in Cartan's paper [9], and for some domains of tube type[7] it occurs in Bochner's paper [5]. Its general theory, independent of the classification and valid for all domains, is due to Korányi and Wolf [54]. It is easily shown that every Siegel domain—it need not be symmetric—is equivalent to a bounded domain. This is the fact which establishes the link with Cartan's problem, because in order to show that not every bounded homogeneous domain is symmetric it now suffices to find a homogeneous Siegel domain which fails to be symmetric. This is what Pjateckiĭ-Šapiro did [65], [66].

Siegel domains are in some sense simpler than bounded domains—they are "more linear". Also they are defined in terms of fairly elementary geometry. I shall give their definition in detail. Some preliminaries are needed for this. A subset Ω of a real vector space V is a cone if (1) $a, b \in \Omega$ implies that $a + b \in \Omega$ and (2) $a \in \Omega, \lambda > 0$ implies $\lambda a \in \Omega$. If Ω is a convex cone its dual Ω' is contained in the dual space V' of V and is defined by $\Omega' = \{x \in V' | \langle x, y \rangle > 0, y \in \Omega\}$ and is also a convex cone. A cone is said to be regular if it is convex, open, and contains no entire straight line (the last condition means that the cone is "pointed"). The importance of regularity resides in the circumstance that the dual cone of a regular cone is regular, and hence in particular open. (See Figure 1.)

Now think of C^n as $C^n = R^n + iR^n$. If A is any subset of R^n

$$T_A = \{z = x + iy \in C^n | y \in A\}$$

[7]These are domains whose half-plane realizations are tube domains. The definition of these follows below.

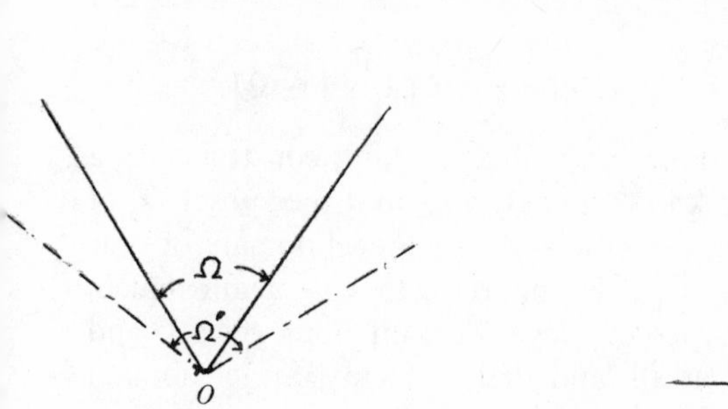

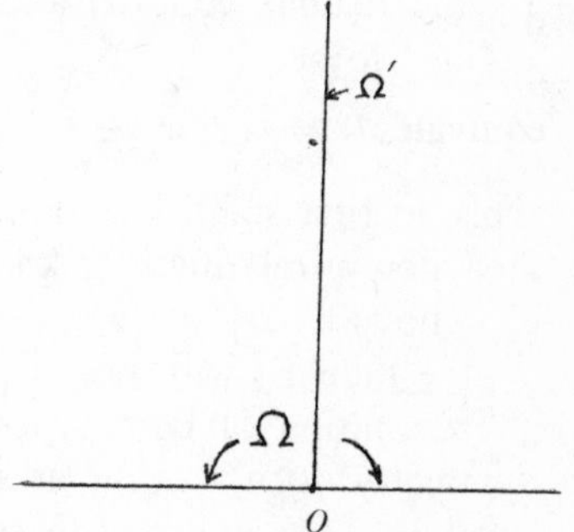

FIG. 1

is called the tube domain with base A (or over A). A Siegel domain of type I is a tube domain over a regular convex cone. The upper half-plane $\mathcal{H}$ in $C^1 : \{z = x + iy | y > 0\}$ is the tube domain over the regular cone $(0, \infty) \subset R^1$. We have encountered earlier another example in connection with the Lie spheres. Tube domains have been considered earlier;[8] it is their role in this theory which is novel. The symmetric Siegel domains of type I were studied in detail by M. Koecher [40] and O. Rothaus [71].

The genuinely new domains are Siegel domains of type II introduced by Pjateckiĭ-Šapiro in [66]. To define them another auxiliary concept is needed. Let Ω be a regular cone in R^n, a mapping $\Phi : C^m \times C^m \to C^n = R^n + iR^n$ is said to be an Ω-positive hermitian map if

(1) $\Phi(u, v)$ is linear in the first argument;
(2) $\Phi(v, u) = \overline{\Phi(u, v)}$ where for $a + ib \in C^n$, $\overline{a + ib} = a - ib$;
(3) $\Phi(u, u) \in \overline{\Omega}$;
(4) $\Phi(u, u) = 0$ if and only if $u = 0$.

The Siegel domain $D = D_{\Omega, \Phi}$ of type II in $C^n \times C^m$, determined

[8]Tube domains with convex open bases provide important and simple examples of domains of holomorphy.

by the regular cone Ω, and the Ω positive hermitian map Φ is defined to be

$$D = \{(z, w) \in C^n \times C^m | \operatorname{Im} z - \Phi(w, w) \in \Omega\}.$$

This, at first sight, is a rather unfamiliar looking construct. As a first step in relating it to known things, note that the subset D_1 of D defined by $D_1 = \{(z, w) \in D | w = 0\}$ is a Siegel domain of type I—the tube T_Ω with base Ω. It is important to note again that in the definition of both types of Siegel domain homogeneity and symmetry do not occur at all, and that in fact, a Siegel domain need not have either of these properties.

A necessary condition for homogeneity of a Siegel domain is that the cone Ω be homogeneous—i.e., that the group of linear transformations of R^n carrying Ω into itself act transitively on Ω. If the domain is of type I this condition is also sufficient. It would lead us too far afield to discuss the sufficient condition for domains of type II. (See [61].) A simple necessary condition for symmetry can also be given. It is again stated in terms of the cone. A regular cone in a real vector space V is said to be self adjoint if V has an inner product such that—after identifying V' by means of the inner product with V—the cone is equal to its dual cone.[9] If a Siegel domain is symmetric, then its cone is self adjoint and homogeneous. As before, for domains of type I this condition is also sufficient. The self adjoint homogeneous cones are easily described (the Lorentz cone is one of them) and their list will accompany that of the symmetric Siegel domains at the end of this section.

Let us now see some concrete examples of Siegel domains of type II. The simplest domain of this type is $D = \{(z, w) \in C \times C | \operatorname{Im} z - |w|^2 > 0\}$. Here the cone is $(0, \infty) \in R^1$, and $\Phi(u, v) = u\bar{v}$. This domain is symmetric, in fact it is equivalent to B^2. It is a very easy calculation to check that the mapping

$$(\zeta_1, \zeta_2) \mapsto \left(i \frac{1 - \zeta_1}{1 + \zeta_1}, \frac{i\zeta_2}{1 + \zeta_1} \right)$$

---

[9]Homogeneous self adjoint regular cones are Riemannian symmetric spaces.

is the Cayley transform which maps B^2 onto D. The second example is the homogeneous Siegel domain of type II in $C^4 = (C^3 \times C^1)$ which was used by Pjateckiĭ-Šapiro in his solution of Cartan's problem. Let $\Omega = \{y \in R^3 | y_1 y_2 - y_3^2 > 0, y_1 > 0\}$. This cone is a circular cone in R^3 as can be readily checked by making the change of coordinates $y_1 = \eta_1 + \eta_2, y_2 = \eta_1 - \eta_2$ (and observing that the defining inequalities of Ω are equivalent to (1) $\eta_1^2 > \eta_2^2 + \eta_3^2$ and $\eta_1 + \eta_2 > 0$, which in turn are equivalent to (1) and $\eta_1 > 0$). The mapping $\Phi : C^1 \times C^1 \to C^3$ is defined by $\Phi(u, v) = (u\bar{v}, 0, 0)$. The domain itself is

$$D = \{(x + iy, w) \in C^3 \times C^1 | (y_1 - |w|^2) y_2 - y_3^2 > 0,$$
$$y_1 - |w|^2 > 0\}.$$

Observe that Ω is self adjoint (as we have seen it is a Lorentz cone), yet it turns out that the domain fails to be symmetric.

In higher dimensions there are even homogeneous Siegel domains of type I which are non-symmetric, because there exist homogeneous convex cones which are not self adjoint. The discovery of this fact is due to E. B. Vinberg [83]. The simplest cone of this nature consists of all pairs

$$\left\{ \begin{bmatrix} a & x \\ x & c \end{bmatrix}, \begin{bmatrix} b & y \\ y & c \end{bmatrix} \right\}$$

of real symmetric, positive definite 2×2 matrices with a common diagonal element c. Its dimension is five.

The role that Siegel domains were to play was not exhausted in the solution of Cartan's problem. In fact, E. B. Vinberg, S. G. Gindikin and I. I. Pjateckiĭ-Šapiro proved [84] in 1963 that *every bounded homogeneous domain is holomorphically equivalent to a homogeneous Siegel domain* (*of type I or II*). This is the fundamental result of the theory, it is the closest analogue of the Riemann mapping theorem that we have in several variables, and it reduces the investigation of bounded homogeneous domains to that of a very special class of domains—the homogeneous Siegel domains. To study the latter means to study the homogeneous cones Ω, the forms Φ, and their interplay. These are clearly circumscribed concrete tasks.

The convex homogeneous cones have been studied intensively by Vinberg [83], and O. Rothaus [72]. Vinberg shows that these cones can be put into one-to-one correspondence with certain generalized matrix algebras—the so called T-algebras. Rothaus has found a method for systematically constructing all homogeneous convex cones from lower dimensional ones. These theories do not, however, provide a "classification" in the narrow sense.[10] All these investigations have a markedly algebraic flavor, and here I have to mention a surprising connection between the theory of homogeneous cones and *Jordan algebras* which was discovered in the fifties by M. Koecher [39], [40]. The methods developed by him and his school are not only important in studying the cones, but also provide a new and independent approach to the theory of bounded symmetric domains. This approach is especially effective and elegant in the case of the domains of tube type, i.e., those whose half-plane realizations are Siegel domains of type I. The application of these algebraic ideas has also led to the latest advance in the subject. In his, as yet unpublished, thesis [14] J. Dorfmeister has obtained a complete structure theory and classification of the homogeneous convex cones.

To conclude the survey of the domains I shall list the half-plane realizations of the Cartan domains. Let me begin by enumerating the self adjoint homogeneous cones.

(a) The $n \times n$ real, symmetric and positive definite matrices.
(b) The $n \times n$ complex, hermitian and positive definite matrices.
(c) The $2n \times 2n$ complex, hermitian and positive definite matrices subject to the condition $Jz = -{}^t zJ$, where ${}^t z$ is the transpose of

---

[10] Ideally, a *classification* of a class of mathematical objects should state that every object of the class can be constructed in an essentially unique way from certain "simple" or "irreducible" objects of the class. A typical example is the structure theorem of finitely generated abelian groups. Cartan's classification of the bounded symmetric domains also is a classification in this sense.

z and $J = \begin{bmatrix} 0 & I \\ -I & 0 \end{bmatrix}$. (This cone can also be thought of as the cone of quaterionic, hermitian and positive definite matrices.)

(d) The spherical or Lorentz-cone in R^n:

$$\left\{ x \in R^n | x_n > +\sqrt{x_1^2 + x_2^2 + \cdots + x_{n-1}^2} \right\}.$$

(e) The set of 3×3 matrices whose entries are Cayley numbers and which are hermitian and positive definite in a suitable sense.

The list of half-plane realizations follows. The letters m and n below denote the same numbers as in the list of bounded domains.

I. $D = \left\{ (z, w) \in M_r \times M_{s,r} | \frac{z - z^*}{2i} - w^*w > 0 \right\}$. The cone is (b); the form is $\Phi(u, v) = v^*u$. The domain is of type I—the tube over the cone (b)—if and only if $s = 0$. The relation between r, s and m, n is $m = r + s$, $n = r$.

II. The domain is the tube over the cone (a) in the space of $n \times n$ complex symmetric matrices.

III. In this case the domains are different depending on the parity of n. Let M_{2r}^J denote the subspace of M_{2r} defined by $Jz = -{}^tzJ$, where $J = \begin{bmatrix} 0 & I \\ -I & 0 \end{bmatrix}$. If $n = 2r$, then the domain is the tube over the cone (c) in M_{2r}^J. If $n = 2r + 1$, then

$$D = \left\{ (z, w) \in M_{2r}^J \times M_{2r,1} | \frac{z - z^*}{2i} - ww^* - (J\bar{w})(J\bar{w})^* > 0 \right\}.$$

The cone is (c) and the form is $\Phi(u, v) = uv^* + (J\bar{u})(J\bar{v})^*$.

IV. The domain is the tube over the Lorentz cone in C^n.

The sixteen dimensional exceptional domain is of type II. Its cone is the eight dimensional Lorentz-cone. The twenty-seven dimensional exceptional domain is a tube over the cone (e).

3. FUNCTION THEORY AND HARMONIC ANALYSIS

We shall be interested principally in the interaction between complex function theory in a domain D of complex n-space and harmonic analysis or real variable theory on its boundary. The example of the polydisc and the torus suggest that it is not the whole boundary which is relevant here but only part of it. The distinguishing property of the torus in relation to the polydisc is that it is a minimal set for which the maximum principle holds. For any bounded symmetric or Siegel domain D in C^n we define the *Šilov boundary*[11] of D as the smallest closed subset $\check{S}$ of the (topological) boundary ∂D with the following property: Every function f which is holomorphic and bounded in D, and continuous on $\bar{D}$ must satisfy $\sup_{z \in D}|f(z)| = \sup_{w \in \check{S}}|f(w)|$.[12] The real dimension of $\check{S}$ is not less than the (complex) dimension of D and it is equal to it if and only if the half-plane realization of D is a Siegel domain of type I—such a D is said to be of *tube type*. If this is the case $\check{S}$ is itself a compact Riemannian symmetric space (it is not in general) and is closely related to the cone Ω of the half-plane realization of D.[13] The dimension of S is equal to $2n - 1$, i.e., $\check{S}$ coincides with the topological boundary of D if and only if D is a ball. Every holomorphic automorphism of D extends to a neighborhood of D, and therefore acts on the boundary of $\check{S}$. The identity component G of the group of automorphisms of D acts transitively on the Šilov boundary, which is the only closed orbit of G in ∂D. Let K denote the subgroup of G which leaves the origin fixed—the stability group of the origin. It is a connected compact group and acts on D (and C^n) by linear transformations, in fact by unitary transformations if the coordinates in C^n are suitably choosen. This group also acts transitively on $\check{S}$ and is the group which is relevant in our context. For the disk K is the circle

---

[11] Also *Bergman-Šilov boundary*, *distinguished boundary*, and in the older literature also *spine*.

[12] If D is bounded symmetric, then continuity on $\bar{D}$ implies boundedness, and the definition coincides with the one given in terms of function algebras.

[13] Ω is the so called non-compact dual of $\check{S}$.

group and acts by multiplication by complex numbers of modulus one. In this case it can be identified with ∂U. The fact that K is compact implies that there is a normalized invariant measure μ on $\check{S}$, i.e., $\mu(\check{S}) = 1$ and

$$\int_{\check{S}} f(kx)\,d\mu(x) = \int_{\check{S}} f(x)\,d\mu(x)$$

for each $k \in K$ and continuous f on $\check{S}$.[14]

Both D and K are *circular sets*—they are carried into themselves by multiplication by complex numbers of modulus one. Further the action of K contains these multiplications. A consequence of this (and of Fubini's theorem) is the following elementary formula

$$\int_{\check{S}}\left(\int_{T} f(\zeta x)\,d\zeta\right)d\mu(x) = \int_{\check{S}} f(x)\,d\mu(x).$$

This is valid for every f which is integrable or measurable and non-negative. Ever since Bochner observed this fact [4] it has been widely used and constitutes an extremely useful artifice. Another important geometric fact is that bounded symmetric domains are convex sets, as R. Hermann showed [60]. This and circularity imply that if $w \in \bar{D}$ then the "slice" $\{z \in C^n | z = \zeta w, \zeta \in U\}$ is contained in D. Bochner's artifice combined with "slicing" reduces many problems of function theory to problems in only one variable. For instance the theory of H^p spaces in its early stages depends very much on this possibility.

For Siegel domains of type I the situation is familiar, being a natural extension of the one prevailing in the case of the upper half-plane. If

$$D \subset R^n + iR^n \quad \text{is defined by} \quad D = \{x + iy | y \in \Omega\},$$

then $\check{S} = R^n$, and the usual group structure of R^n is the one relevant for harmonic analysis. For Siegel domains of Type II things are more involved. If

$$D = D_{\Omega,\Phi} = \{(z, w) \in C^n \times C^m | \operatorname{Im} z - \Phi(w, w) \in \Omega\},$$

---

[14]The relation then extends to $L^1(\check{S})$.

then

$$\check{S} = \{(z, w) \in C^n \times C^m | \text{Im } z - \Phi(w, w) = 0\}.$$

If one writes z as $z = x + iy$, then on $\check{S}$ $y = \Phi(w, w)$, and $\check{S}$ is parametrized by a point x in R^n and a point w in C^m, and $\check{S}$ is homeomorphic to $R^n \times C^m$. It is frequently useful to write $t = y - \Phi(w, w)$ and to use the coordinates $x \in R^n$, $t \in \Omega$, and $w \in C^n$. In this coordinate system $\check{S}$ is simply the set of (x, t, w)'s for which $t = 0$. Whenever a t occurs in a formula relating to $\check{S}$, it is a warning that this description of $\check{S}$ is being used. If the context allows I shall denote points (x, w) of $\check{S}$ by a single letter with a dot. The set $\check{S}$ also carries a group structure.[15] This group is not commutative and is a nilpotent Lie group. The description of $\check{S}$ by the coordinates (x, w) is particularly convenient whenever group properties have to be discussed. The multiplication of the elements (x, w) and (x', w') is given by

$$(x, w) \cdot (x', w') = (x + x' + 2 \text{ Im } C(w, w'), w + w').$$

Also $(x, w)^{-1} = (-x, -w)$. The center of the group consists of the elements $(x, 0)$, and it is an easy and useful exercise to show that the commutator of two elements already lies in the center—the group is nilpotent of step two. This means that it is as little non-commutative as it possibly can be. The group $\check{S}$ acts on the domain D by affine transformations as follows

$$(\xi, \omega) \cdot (x + iy, w) = (x + \xi + 2i\Phi(w, \omega) + i\Phi(\omega, \omega), w + \omega).$$

If the domain is of type I this reduces to the obvious action of R^n on T_Ω given by

$$\xi \cdot (x + iy) = x + \xi + iy.$$

[15]To be more precise there is a group N of affine transformations of D (even if D is not symmetric) which acts simply transitively on $\check{S}$—i.e., it is transitive and $g \cdot x = h \cdot x$ implies $g = h$: therefore N can be identified with $\check{S}$. Here I shall only consider $\check{S}$ and ignore N.

The fact that the Šilov boundaries of Siegel domains actually are groups—not only homogeneous spaces of groups as the Šilov boundaries of the Cartan domains—makes the analogy between these domains and the upper half-plane closer than the corresponding analogy between the disc and the Cartan domains. The Haar measure on $\check{S}$ is just Lebesgue measure on $R^n \times C^m$, and will be demoted by $d\beta$. Siegel domains of both types are convex. (This is obvious for type I; for type II the proof is an exercise in calculating with Φ.) However, they are clearly not circular; so there is no analogue of the method based on Bochner's artifice and slicing.

Two of the most important tools of analysis on the unit disc and the upper half-plane are the Poisson and the Cauchy integrals. If $0 \leqslant r < 1$ and $z = re^{i\theta}$, then the Poisson integral of a function f, square integrable on $T = \partial U$, is

$$\int_0^{2\pi} \frac{1 - r^2}{1 + r^2 - 2r\cos(\theta - \phi)} f(e^{i\phi}) \frac{d\phi}{2\pi}$$

$$= \int_0^{2\pi} \frac{1 - |z|^2}{|1 - ze^{-i\phi}|^2} f(e^{i\phi}) \frac{d\phi}{2\pi},$$

and

$$P(z, e^{i\phi}) = P_z(e^{i\phi}) = P_r(e^{i(\theta-\phi)})$$

$$= \frac{1 - r^2}{1 + r^2 - 2r\cos(\theta - \phi)} = \frac{1 - |z|^2}{|1 - ze^{-i\phi}|^2}$$

is the Poisson kernel. Let z and f be as above. Then the Cauchy integral of f on ∂U is

$$\frac{1}{2\pi} \int_{\partial U} \frac{f(\zeta)}{\zeta - z} d\zeta = \int_0^{2\pi} \frac{f(e^{i\phi})}{1 - ze^{-i\phi}} \frac{d\phi}{2\pi} = (f, S_z)$$

where the inner product is taken in L^2 of T (relative to the Haar measure $d\phi/2\pi$ of T), and

$$S(z, w) = S_z(w) = \frac{1}{1 - \bar{z}w}.$$

Observe that S is a holomorphic function of w and an antiholomorphic function of z. If f is holomorphic in a neighborhood of $\bar{U}$, then

$$f(z) = (f, S_z).$$

The kernel of Cauchy's integral thus specialized to the boundary of U is also called the Szegö kernel, and the integral (f, S_z) itself is referred to as the Szegö integral. I shall use both terms interchangeably. There is a third kernel which is of essential importance for the geometry of the domains, the Bergman kernel, but in this exposition it can be ignored without serious loss.

Let us now think of various ways of interpreting these formulas which might lend themselves to extension to higher dimensions. The Poisson kernel has several interpretations:

1. It is the kernel associated with Abel summation of Fourier series.
2. It is the normal derivative of Green's function for the Dirichlet problem on U.
3. Formally it is related to the Cauchy kernel by

$$\frac{1}{1 - \bar{z}e^{i\phi}} = \frac{1}{2}\left(\frac{1 - |z|^2}{|1 - \bar{z}e^{i\phi}|^2} + 1 - i\,\frac{2\operatorname{Im}(ze^{-i\phi})}{|1 - \bar{z}e^{i\phi}|^2}\right).$$

   The imaginary part on the right hand side is the conjugate Poisson kernel.

4. A less familiar relation between S and P is

$$P(z, e^{i\phi}) = \frac{|S(z, e^{i\phi})|^2}{S(z, z)}.$$

5. Finally the most unfamiliar way of looking at the Poisson kernel is the following one. Let $g : z \mapsto e^{i\alpha}(z - z_0)/(1 - \bar{z}_0 z)$

be an automorphism of U. It extends to a neighborhood of U and its restriction to ∂U is a diffeomorphism of ∂U. It is an amusing calculation to verify that $P(g(0), \cdot)$ is the absolute value of the Jacobian of the transformation g^{-1} of ∂U. Concretely if f is continuous on ∂U, then

$$\int_0^{2\pi} f(g(e^{i\phi}))\frac{d\phi}{2\pi} = \int_0^{2\pi} f(e^{i\phi})P(g(0), e^{i\phi})\frac{d\phi}{2\pi}.$$

It turns out that it is the last two of the above ways of viewing the Poisson kernel which lend themselves to generalization. It would be interesting to know whether 2. has useful analogues in higher dimensions. For some domains, results in this direction were obtained by J. Mitchell [59].

It seems less easy to think of alternative ways of looking at the Cauchy kernel. The clue here is given by the unusual way of writing the Cauchy formula as $f(z) = (f, S_z)$ combined with some elementary facts from the theory of the Hardy space H^2: This provides the "right" way of viewing S_z. The function $t \mapsto S_z(t)$ is continuous for fixed $z \in U$ and hence an element of H^2. It is not hard to see that for $f \in H^2$ we still have

$$f(z) = (f, S_z).$$

This means S_z is the reproducing kernel[16] of H^2, i.e., an element of H^2 depending on a parameter $z \in U$, such that the linear functional determined by it is evaluation at the point z. Once one has the Szegö kernel, the Poisson integral can be defined by 4. This method works both for bounded symmetric domains and Siegel domains, symmetric or not. The fifth interpretation of the Poisson integral also leads to a successful generalization but is restricted to the symmetric case. We shall see that where both definitions are available they coincide. Since the method of widest scope to generalize the kernel functions depends on determining the repro-

[16]Reproducing kernels on the boundary were first considered by G. Szegö in connection with orthogonal polynomials. This is the origin of the name for S_z.

ducing kernel of H^2, I shall begin the survey of function theory proper by discussing the space H^2. Since the definition is the same for H^p with any p I shall give that, and occasionally I shall insert a comment on $p \neq 2$. A further account of these spaces comes at the end of the article.

If f is a function defined on a bounded symmetric domain D and $0 \leqslant r < 1$ then $f_r : \check{S} \to C$ is defined by $F_r(\dot{z}) = f(r\dot{z})$. Let now $0 < p < \infty$, then a function f holomorphic in D is said to belong to the (Hardy) space $H^p(D)$ if

$$\sup_{0 \leqslant r < 1} \int_{\check{S}} |f_r|^p d\mu < \infty.$$

$H^\infty(D)$ is the Banach space of bounded holomorphic functions on D, with the norm $\|f\|_\infty = \sup_{z \in D} |f(z)|$. One can also define the Nevanlinna space $N(D)$, which consists of the functions which are holomorphic in D and which satisfy the condition

$$\sup_{0 \leqslant r < 1} \int_{\check{S}} \log^+ |f_r| d\mu < \infty, \qquad \log^+ x = \begin{cases} \log x \text{ if } x \geqslant 1, \\ 0 \text{ if } 0 \leqslant x < 1. \end{cases}$$

N contains all the spaces H^p. The spaces H^p and N are linear spaces and for $1 \leqslant p$ the pth root of the supremum in the inequality defining H^p is easily seen to be a norm. Using Bochner's artifice and the one variable theory, one can quite easily show that for $f \in N$, $f_r(\dot{z})$ tends to a limit $f^*(\dot{z})$ for almost every $\dot{z} \in \check{S}$ and that the function f^* thus defined on $\check{S}$, the boundary function of f, belongs to $L^p(\check{S})$ if f belongs to H^p. I shall denote the space of all such f^* by $H^p(\check{S})$. It also follows without difficulty that for r tending to one

$$\int_{\check{S}} |f_r - f^*|^p d\mu \to 0,$$

and that

$$\sup_{0 \leqslant r < 1} \int |f_r|^p d\mu = \int_{\check{S}} |f^*|^p d\mu.$$

The mapping $f \mapsto f^*$ is an isomorphism of $H^p(D)$ onto $H^p(\check{S})$ for

$0 < p < \infty$. For $0 < p < 1$, H^p can be made into a metric space too, by defining the distance between f and g by $(\int |f^* - g^*|^p \, d\mu)$. All the results so far are quite simple thanks to Bochner's formula.[17] The proof that the space of boundary functions $H^p(\check{S})$ is closed in $L^p(\check{S})$, and that therefore H^p is a complete metric space is more delicate. The completeness of these spaces is equivalent to the fact that they are equal to those obtained by closing the polynomials relative to the L^p metric on the Šilov boundary. The spaces H^p are studied in some detail by K. T. Hahn and J. Mitchell in [23] and [24], the space N by M. Stoll [81].

Let us turn now to the Siegel domains. If f is a function defined on D and $t \in \Omega$, then $f_t : \check{S} \to C$ is defined by

$$f_t(\dot{z}) = f(\dot{z} + it), \text{ if } D \text{ is of type I, and by}$$

$$f_t(\dot{z}, \dot{w}) = f(\dot{z} + it, \dot{w}), \text{ if } D \text{ is of type II.}$$

For $0 < p < \infty$ $H^p(D)$ is defined as the set of functions holomorphic on D such that

$$\sup_{t \in \Omega} \int_{\check{S}} |f_t|^p \, d\beta < \infty.$$

$H^\infty(D)$ is defined as before. The above definition is more complicated than the one for bounded domains because the supremum in the inequality defining H^p is taken not over a line segment but over a set which has half the (real) dimension of D. The half-planes are not circular, and the reduction to the one variable case no longer is possible. This makes the proof of the elementary results described above more complicated and sets the theory for $p = 2$ apart from that for $p \neq 2$. Now the formal definition of the Szegö kernel can be given. Let D be either a bounded symmetric domain or a Siegel domain. A function $z \mapsto S_z$ which maps D into $H^2(\check{S})$ is the Szegö kernel of $H^2(D)$ if for every $f \in H^2(D)$,

---

[17]In fact Bochner has proved the above results for a larger class of bounded circular domains—those on which the analogue of his artifice (see p. 273) holds.

$f(z) = (f^*, S_z) = \int_{\check{S}} f^*(u)\overline{S_z(u)}\, du$. Here $du = d\mu(u)$ if D is a bounded symmetric domain, and $du = d\beta(u)$ if D is a Siegel domain. The Szegö kernel is unique if it exists—we shall see that this is always the case.

The H^2 spaces for Siegel domains of type I were first studied in 1944 by Bochner in his paper *Group invariance of Cauchy's formula in several variables* [5]. Harmonic analysis on symmetric domains begins with this paper. Let then T_Ω be a Siegel domain of type I. Bochner shows that a function f belongs to $H^2(T_\Omega)$ if and only if

$$f(z) = \int_{\Omega'} e^{2\pi i \langle \lambda, z \rangle} g(\lambda)\, d\lambda$$

with $g \in L^2(\Omega')$. From this it easily follows that f has a boundary function f^* on $\check{S}$ such that

$$\int_{\check{S}} |f^*(x) - f(x+iy)|^2 dx \to 0,$$

as y tends to 0 in Ω.

The L^2 Fourier transform of the boundary function f^* is equal to $g(\lambda)$ for $\lambda \in \Omega'$, and zero almost everywhere outside Ω'. The representation formula therefore can be written as

$$\begin{aligned} f(z) &= \int_{\Omega'} e^{2\pi i \langle \lambda, z \rangle} g(\lambda)\, d\lambda \\ &= \int_{\check{S}} e^{2\pi i \langle \lambda, z \rangle} \hat{f}^*(\lambda)\, d\lambda. \end{aligned}$$

This formula gives a characterization of H^2 functions by means of the Fourier transforms of their boundary functions. I shall return to this later. Here I want to concentrate on the role of H^2 in the construction of the kernel functions. It immediately follows from the above that $H^2(\check{S})$ is a closed subspace of $L^2(\check{S})$, and that therefore $H^2(\check{S})$ and $H^2(D)$ are Hilbert spaces.

Bochner then considers the function

$$K(z) = \int_{\Omega'} e^{2\pi i\langle\lambda, z\rangle} d\lambda, \qquad z \in T_\Omega$$

which is holomorphic in T_Ω and shows that for $f \in H^2(T_\Omega)$

$$f(x + iy) = \int_{\check{S}} f^*(s) K(x - s + iy)\, ds$$

$$= \int_{\check{S}} f^*(s) K_y(x - s) = (f^* * K_y)(x).$$

The kernel K is essentially the Szegö kernel of $T_\Omega : S_z(s) = K(z - s)$.

Bochner then proceeds to calculate K explicitly for most of those T_Ω which are half-plane realizations of classical domains, and then using the Cayley transform also finds explicit formulas for the Szegö kernel on some of the classical domains. The Szegö kernel for all the (bounded) classical domains (not only for those of tube type) were determined by Hua [28]. He does not mention H^2 but considers the closure of the polynomials relative to the L^2 norm on $\check{S}$—we have seen that the two spaces are equal; constructs the Szegö and Bergman kernels from an orthonormal basis of this space; and defines the Poisson kernel by 4 (see page 276) and establishes some of its basic properties.

For Siegel domains of type II the basic results about H^2 are due to S. G. Gindikin [21]. He proved that for $f \in H^2(D)$, f_t converges in $L^2(\check{S})$ norm to an element $f^* \in L^2(\check{S})$ as $t \in \Omega$ tends to zero, and that the space $H^2(\check{S})$ of boundary functions is a Hilbert space. He also determined the Szegö (and Bergman) kernels, and gave a representation formula for functions of H^2:

$$f(z, w) = \int_{\Omega'} e^{2\pi i\langle\lambda, z\rangle} g(\lambda, w)\, d\lambda.$$

These results are much more difficult to prove than those for tube domains. In fact, Gindikin's arguments are not conclusive, and a complete proof of his results was given only recently by A.

Korányi and E. M. Stein who also found a new representation formula [49]. The spaces H^p on tube domains for $p \neq 2$ were first considered by E. M. Stein, G. Weiss, and M. Weiss [79], and on Siegel domains of type II by A. Korányi [44], and E. M. Stein [76]. A fairly large part of the theory of these spaces—especially if $p < 1$—deals with proving the existence of boundary values and I shall return to it at the end of the article when I discuss the boundary behavior of holomorphic functions.

Once the existence of the Szegö kernel is secured, one defines the Poisson kernel by

$$P_z(\dot{u}) = \frac{|S_z(\dot{u})|^2}{S_z(z)} \qquad z \in D, \quad \dot{u} \in \check{S}.$$

Korányi [42] and Stein et al. [79] did this for domains of type I, and established the basic properties of the kernel—mainly that it is an approximate identity. For domains of type II this was done by Korányi who then used the general theory of the Cayley transform developed by him and J. A. Wolf to determine the Szegö and Poisson kernels for all bounded symmetric domains [44].

Now, as I hinted at the beginning of this section, on the symmetric domains the Poisson kernel can be generalized in a different way. Recall the last interpretation of the Poisson integral given on page 276, namely that on the disc the Poisson kernel can be regarded as the Jacobian of an automorphism restricted to the boundary. This observation is the starting point of a group theoretic approach to the definition of the Poisson kernel and Poisson integral. For the bounded symmetric domains this approach seems to have been first used systematically by D. Lowdenslager [56]. It was shown to work on arbitrary non-compact Riemannian symmetric spaces by H. Furstenberg [19]. In fact these spaces constitute the natural setting for the group theoretic approach.[18] The question now arises whether these two possible definitions of the

[18]A very clear and readable exposition of this method can be found in Furstenberg's article [20].

Poisson kernel on a symmetric domain coincide. The answer is yes [44], and depends on *harmonicity* which now finally makes its appearance.

For the one variable case the kernel $P_z(\dot{u})$ as a function of z is harmonic. On every Siegel domain one can define a second order elliptic partial differential operator in terms of the Bergman metric —the so-called Laplace-Beltrami operator Δ. The harmonic functions on D are defined to be those which satisfy $\Delta u = 0$. For non-symmetric Siegel domains it is not known whether the Poisson kernel is harmonic in this sense. (The Szegö kernel is, due to its holomorphy.) If D is symmetric then there are several notions of harmonicity. The most restrictive one—strong harmonicity—is defined in terms of the group of automorphisms of D. If u is strongly harmonic, then $\Delta u = 0$, but the converse is not true in general. Korányi showed that if D is symmetric then $z \to P_z(u)$, and therefore Poisson integrals are strongly harmonic. As long as we don't know whether the Poisson kernel is harmonic on non-symmetric Siegel domains, the natural context for the study of harmonic functions and their representability as Poisson integrals is that of symmetric spaces. The results valid for the symmetric domains follow from these more general ones. I shall not enter at all into the discussion of what harmonic functions are representable as Poisson integrals.[19] To make amends for such cavalier treatment of an interesting subject, let me describe one more result about harmonic functions. (The proof of the harmonicity of P for symmetric D depends on it.) This result is a remarkable generalization of Gauss's mean value theorem due to R. Godement [22]. Gauss's theorem states that a continuous function u on C is harmonic if and only if for all $z \in C$

$$u(z_0) = \int_0^{2\pi} u\big(z_0 + e^{i\phi}(z - z_0)\big)\, \frac{d\phi}{2\pi}.$$

[19]Harmonic functions on symmetric spaces, and in particular their representations by Poisson integrals is discussed in Korányi's article [47].

(This integral is the mean value of u on the circle of radius $|z - z_0|$ centered at z_0, written in form convenient for the purpose at hand.) The affine transformations of the plane which are isometries (of the Euclidean metric of $C = R^2$) are of the form $z \mapsto a + e^{i\phi}z$ with ϕ real, a complex. These transformations form a group of which the plane is a homogeneous space. The stability group K_{z_0} of the point z_0 is the set of affine transformations k such that $kz = z_0 + e^{i\phi}(z - z_0)$. It is isomorphic to T. The mean value formula now can be written as follows:

$$u(z_0) = \int_{K_{z_0}} u(kz)dk.$$

In this form the integral is meaningful if z_0 is a point of a symmetric domain D with K_{z_0} its stability group, and integration is relative to the Haar measure of K_{z_0}. Godement's theorem states that a continuous function on D is strongly harmonic if and only if the above formula holds for all z in D.

Let us return to the Poisson integral on Siegel domains. I will limit myself to dealing with these; but similar results hold for the bounded symmetric domains. The kernel P_z belongs to L^p for $1 \leqslant p \leqslant \infty$. It is an approximate identity, and it reproduces H^p functions for $1 \leqslant p \leqslant \infty$—i.e.,

$$f(z) = \int_{\dot{S}} P_z(\dot{u}) f^*(\dot{u}) d\beta(\dot{u}).$$

The kernel for $\mathcal{H}^n$, $n \geqslant 1$, has a simple and important semigroup property (for $\mathcal{H}^1 = \mathcal{H}$, see page 267). For $n = 1$ it is

$$\int_R P_y(x - s)P_t(s)ds = P_{y+t}(x).$$

As a consequence, if U_t is the Poisson integral of u, then

$$\int_R U_t(s)P_y(x - s)ds = U_{t+y}(x).$$

Both the above identities fail to be true if $D \neq \mathcal{H}^n$. This is the source of many difficulties: e.g., if D is a symmetric Siegel domain and U a harmonic function on D—say the Poisson integral of an L^1 function on $\check{S}$—then $z \mapsto U_t(z) = U(z + t)$, $t \in \Omega$, is not harmonic. The second identity however remains valid if u is the boundary function of a holomorphic function. In particular for $f \in H^2(\check{S})$ the Poisson integral of f equals its Szegö integral.

Another important fact is that in the one variable theory, and on $\mathcal{H}^n$ the Cauchy and Poisson integrals can be written as convolutions. It is practically obvious that for Siegel domains of type I this continues to be true. For domains of type II it is also true, but the non-commutativity of the group $\check{S}$ introduces unfamiliar elements—e.g., there is more than one way to define convolution, which no longer is commutative. Therefore let me give a few words of explanation. The crucial property of the kernels in this context is their group invariance relative to $\check{S}$. Let $\zeta \in D$, $\dot{w} \in \check{S}$ and $\dot{\alpha} \in \check{S}$, and recall that $\check{S}$ acts on D. Then

$$S(\dot{\alpha}\zeta, \dot{\alpha}\dot{w}) = S(\zeta, \dot{w}),$$

$$P(\dot{\alpha}\zeta, \dot{\alpha}\dot{w}) = P(\zeta, \dot{w}).$$

Let us recast these formulas in more convenient form: Let $\zeta = \dot{\xi} + (it, 0)$ with $\dot{\xi} \in \check{S}$ and $t \in \Omega$, then the kernels can be rewritten as $S(\zeta, \dot{w}) = S_t(\dot{\xi}, \dot{w})$, and $P(\zeta, \dot{w}) = P_t(\dot{\xi}, \dot{w})$. The functions S_t and P_t are now functions on $\check{S} \times \check{S}$. The above formulas expressing the invariance of the kernels become

$$S_t(\dot{\alpha}\dot{\xi}, \dot{\alpha}\dot{w}) = S_t(\dot{\xi}, \dot{w}),$$

$$P_t(\dot{\alpha}\dot{\xi}, \dot{\alpha}\dot{w}) = P_t(\dot{\xi}, \dot{w}).$$

We can now show that the Poisson and Szegö integrals can be written as convolutions. Let then $u \mapsto Tu$ be a linear mapping of the space $L^2(\check{S})$ into itself which is an integral operator—$(Tu)(\dot{\xi}) = \int_{\check{S}} G(\dot{\xi}, \dot{\sigma})u(\dot{\sigma})\, d\beta(\dot{\sigma})$—whose kernel is continuous and has the property that $G(\dot{\xi}, \cdot)$ belongs to L^2 or L^1 for every $\dot{\xi} \in \check{S}$. (P_t and S_t have these properties.) Suppose that G is invariant, i.e., that $G(\dot{\alpha}\dot{\xi}, \dot{\alpha}\dot{\sigma}) = G(\dot{\xi}, \dot{\sigma})$ for all $\dot{\alpha}, \dot{\xi}, \dot{\sigma}$ in $\check{S}$. Apply invariance with $\dot{\alpha} = \dot{\sigma}^{-1}$ to see that $G(\dot{\xi}, \dot{\sigma}) = G(\dot{\sigma}^{-1}\dot{\xi}, 0)$. Define a new kernel K

by $K(\dot{\xi}) = G(\dot{\xi}, 0)$. Then

$$(Tu)(\dot{\xi}) = \int_{\check{S}} K(\dot{\sigma}^{-1}\dot{\xi})u(\dot{\sigma})d\beta(\dot{\sigma}).$$

The integral in this formula is written $u*K$ and is one of the convolutions of $\check{S}$. (Note that $u*K \neq K*u$.) The fact that Tu is the convolution of u with K has an important consequence. For $\dot{\alpha} \in \check{S}$ and $u \in L^2(S)$, define $L_{\dot{\alpha}}u$—the left translate of u by $\dot{\alpha}$—as $(L_{\dot{\alpha}}u)(\dot{\xi}) = u(\dot{\alpha}^{-1}\dot{\xi})$. ($L_{\dot{\alpha}}$ is a unitary operator on L^2.) A simple change of variable easily verifies that $L_{\dot{\alpha}}T = TL_{\dot{\alpha}}$ – i.e., that T is invariant relative to left translations by elements of $\check{S}$.

The basic tools of "analysis on the boundary" are now constructed. The proper setting for this construction and for the study of their elementary properties was the space H^2. Before looking at the more advanced uses to which these tools can be put, let us briefly take up a topic whose treatment I deferred in order to concentrate on the kernel functions—the characterization of H^2 by Fourier analysis.[20] A function f belonging to $L^2(T)$ is (the boundary function of) an element of $H^2(U)$ if and only if its Fourier coefficients

$$\hat{f}_n = \int_0^{2\pi} e^{-in\theta} f(e^{i\theta}) \frac{d\theta}{2\pi}$$

vanish for $n < 0$. This is an immediate consequence of the definitions, and in fact sometimes H^2 is defined by this equivalent property. The analogous theorem for bounded symmetric domains, due to W. Schmid, is fairly recent and quite difficult [74]. Again, let K denote the stability group of the origin in the identity component of the group of holomorphic automorphisms of the domain D. The group K is compact and acts transitively on the Šilov boundary $\check{S}$ of D. (Recall that if $D = U$—the unit disc—then $K = T$ and can be actually identified with the Šilov boundary of U which in this case is all of the boundary.) Schmid

[20] The results for H^p, $p \neq 2$ and $1 \leqslant p \leqslant \infty$, are much less complete.

characterizes those irreducible representations of K which occur in the Fourier decomposition of a function belonging to $H^2(\check{S})$.[21] Some of these results for the classical domains are implicit—in a not altogether obvious way—in earlier work of Hua [28]. In his paper Schmid proved another interesting fact (which was known earlier in the case of the ball): If D contains irreducible factors which are not of tube type, then a function in $H^2(\check{S})$ has to satisfy the so-called tangential Cauchy-Riemann equations on $\check{S}$ .In fact, if D contains no irreducible factor of tube type, then this condition is also sufficient for membership in $H^2(\check{S})$.

For the upper half-plane the characterization of H^2 is given in a well-known theorem of Payley and Wiener which states that $f \in L^2(R)$ is an element of $H^2(R)$ if and only if its Fourier transform $\hat{f}(\xi)$ vanishes for almost all $\xi < 0$. This theorem was extended to Siegel domains of type I by Bochner [5]. We have already seen Bochner's representation formula for H^2 functions:

$$f_t(x) = \int_{\Omega'} e^{2\pi i\langle x, \lambda\rangle} e^{-2\pi\langle\lambda, t\rangle} \hat{f}^*(\lambda)\, d\lambda, \qquad t \in \Omega.$$

If we let t tend to zero in this expression, we obtain the exact analogue of the Payley-Wiener characterization of H^2: A function $f \in L^2(\check{S})$ belongs to $H^2(\check{S})$ if and only if its Fourier transform vanishes almost everywhere outside Ω'. Another fact which follows easily is that for $f \in L^2(\check{S})$ and $t \in \Omega$

$$(K_t * f)\hat{}\,(\lambda) = e^{-2\pi\langle\lambda, t\rangle} \chi_{\Omega'}(\lambda) \hat{f}(\lambda)$$

where $\chi_{\Omega'}$ is the characteristic function of Ω'. If we denote the orthogonal projection of $L^2(\check{S})$ onto $H^2(\check{S})$ by P, then the last formula immediately implies

$$(Pf)\hat{} = \chi_{\Omega'} \hat{f} = \left(\lim_{t \to 0} K_t * f\right)\hat{}\,.$$

---

[21]For information about the harmonic analysis on compact groups the reader is referred to the article of Guido Weiss in this volume.

If we define $\overline{H}^2(\check{S})$ by $\overline{H}^2(\check{S}) = \{g \mid \bar{g} \in H^2(\check{S})\}$, then there is a completely symmetric theory of this space, with the only difference that Ω and Ω' are replaced by $-\Omega$ and $-\Omega'$. In particular the orthogonal projection onto $\overline{H}^2$ is given by

$$(\overline{P}f)\hat{} = \chi_{-\Omega'}\hat{f} = \left(\lim_{t\to 0} \overline{K}_t * f\right)\hat{}.$$

From this it follows immediately that H^2 is orthogonal to $\overline{H}^2$. If $T_\Omega = \mathcal{H}^1$, then $\Omega' \cup (-\Omega') \cup \{0\} = R$, and $\chi_\Omega + \chi_{\Omega'} = 1$ a.e., in other words $P + P = I$. This expresses the familiar fact that H^2 and $\overline{H}^2$ together exhaust L^2, i.e., $L^2 = H^2 \oplus \overline{H}^2$ (orthogonal direct sum). In higher dimensions this no longer is true because the complement of $\Omega' \cup (-\Omega')$ is a set in the nonempty interior in R^n if $n \geqslant 2$. Therefore in this case $L^2(\check{S}) = H^2(\check{S}) \oplus \overline{H}^2(\check{S}) \oplus M$, where M is a non-zero closed subspace of $L^2(\check{S})$ whose elements are, loosely speaking, neither holomorphic nor antiholomorphic.

For Siegel domains of type II Gindikin's representation formula

$$f(z, \zeta) = \int_{\Omega'} e^{2\pi i\langle\lambda, z\rangle} g(\lambda, \zeta)\, d\lambda$$

gives a necessary condition for a function $f \in L^2(\check{S})$ to belong to H^2. From the above formula it follows easily that

$$\int_{R^n} e^{-2\pi i\langle\lambda, x\rangle} f(x, \zeta)\, dx$$

must vanish almost everywhere outside of Ω' for almost all $\zeta \in C^m$. It is shown in [82] that this condition together with the further one that f satisfy the tangential Cauchy-Riemann equations on $\check{S}$ is sufficient for membership in $H^2(\check{S})$. It is proved in the same paper that if the domain is symmetric and has no irreducible factor of type I, then the second condition by itself is sufficient. The Šilov boundary of a Siegel domain of type II is a nilpotent Lie group. Nilpotent Lie groups and their harmonic analysis are fairly well understood. A self-contained exposition of their theory can be found in L. Pukánszky's book [69] which also gives references to original papers. It is natural to ask whether the harmonic analysis of the group $\check{S}$ has connections with the function theory of the domain. This question has been answered in the affirmative by R. D. Ogden and myself: The irreducible unitary representations[22] of $\check{S}$ which occur in the decomposition of an H^2 function can be explicitly characterized. It also can be shown that the representation theorems of Gindikin and Korányi-Stein mentioned earlier are special cases of the "Fourier inversion formula" for the group $\check{S}$. The results were announced in [64].

To conclude this section, let me mention that groups of the type of $\check{S}$ have interesting connections with other areas of mathematics. The simplest of these groups, the one corresponding to the half-plane realization of B^2—it is the prototype of the others—is frequently called the Heisenberg group because it plays a role in the quantum mechanical commutation relations [63]. It has a very simple realization as a matrix group; it is the set of 3×3 real matrices of the form

$$\begin{bmatrix} 1 & a & c \\ 0 & 1 & b \\ 0 & 0 & 1 \end{bmatrix}.$$

(The mapping

$$\begin{bmatrix} 1 & a & c \\ 0 & 1 & b \\ 0 & 0 & 1 \end{bmatrix} \to \left(c - \frac{ab}{2}, \frac{b + ia}{2} \right)$$

is easily seen to be an isomorphism.) These groups also play a role in the theory of Theta functions [2, 11, 30, 48], and quite recently G. B. Folland and E. M. Stein [18] applied them to the study of the so-called $\bar{\partial}$—boundary complex in the theory of partial differential equations.[23]

---

[22]For nilpotent Lie groups (which are not commutative) all irreducible unitary representations are either one dimsional or infinite dimensional, and the Fourier transform has to be defined in operator theoretic terms. Paul Sally's article in this volume discusses infinite dimensional group representation theory.

[23]See E. M. Stein's article in this volume.

The chapter of classical harmonic analysis which deals with the conjugate function operator and the Hilbert transform is one in which real and complex variable methods interpenetrate one another extremely fruitfully. Let us now see what shape this theory takes in several variables. I shall first briefly review the one variable theory in a form suitable to compare it with its extension to Siegel domains. Analogous considerations hold, of course, for bounded domains.

The basic problem can be stated as follows: Is convolution with the Szegö kernel a bounded operator on L^p, $1 < p < \infty$? The Szegö kernel (as a convolution kernel) for $\mathcal{H}$ is given by

$$K_y(x) = \overline{S(x+iy,0)} = \frac{1}{-2\pi i} \cdot \frac{1}{x+iy}$$

$$= \tfrac{1}{2}\left\{ \frac{1}{\pi} \cdot \frac{y}{x^2+y^2} + \frac{i}{\pi} \cdot \frac{x}{x^2+y^2} \right\}$$

$$= \tfrac{1}{2}\{P_y(x) + iQ_y(x)\}$$

when P_y and Q_y are the Poisson and the conjugate Poisson kernels respectively. Recall that (1) $P_t \in L^p$, $1 \leqslant p \leqslant \infty$, (2) $P_t > 0$, and (3) P_t is an approximate identity. Therefore as a convolution kernel P_t is extremely well behaved, and the study of K_y as a convolution kernel is equivalent to that of Q_y. This explains why in the one variable theory the operator $f \mapsto Q_y * f$ plays a dominant role, while K_y itself remains in the background. Note that Q_y is not in L^1 (although in L^p for $1 < p < \infty$), and that it is not positive. Also note that for $y = 0$ and $x \neq 0$ $K_0(x) = (i/2)Q_0(x) = (i/2\pi)(1/x)$. The answer to the question raised at the beginning of this paragraph is affirmative and follows from the famous theorem of M. Riesz [70; 91, I, p. 253].

THEOREM Q: *If $f \in L^p(R)$, $1 < p < \infty$, then $\|Q_y * f\|_p \leqslant A_p \|f\|_p$, where A_p is a constant independent of f and y.*

In view of $K_y = \frac{1}{2}(P_y + iQ_y)$ and $P_y \in L^1$, Theorem Q is equivalent to

THEOREM K: *If* $f \in L^p(R)$, $1 < p < \infty$, *then* $\|K_y * f\|_p \leqslant A_p\|f\|_p$ *where* A_p *is a constant independent of f and y.*

This is the answer to our initial question. Theorem Q has the following obvious corollary about conjugate functions:

THEOREM C: *If* $f = u + iv \in H^p(x)$, $1 < p < \infty$, *then* $\|v_y\|_p \leqslant A_p\|u_y\|_p$ *and* A_p *does not depend on f or y.*

Observe that Q says more than C: namely, that every real element of $L^p(R)$, $1 < p < \infty$, is the real part of an H^p function.

Combined with C, this fact implies Q. Further it is in turn equivalent to the fact that H^p and $\overline{H^p}$—the set of complex conjugates of H^p functions—exhaust $L^p(R)$. This means that every $f \in L^p$ is uniquely a sum $f = g + \bar{h}$ with g and h in H^p.[24] Finally, observe that the mapping $u \mapsto v$ in Theorem C can be defined without reference to any kernel.

Let us now ask further: What happens if in $f * K_y$ we let y tend to zero? What are the meanings of $\lim_{y\to 0} f * K_y$ and $f * K_0$? How are they related, and in particular is the former equal to f if $f \in H^p$? The first thing to note is that $f * K_0$ as it stands has no meaning because $K_0(x) = (i/2\pi)(1/x)$ is not integrable. The link between the boundary operator and the operator with $y \neq 0$ is established by the following limit formula:[25]

$$\text{(L)} \qquad \int_R K_y(x-s)f(s)dx - \int_{|x-s|>y} K_0(x-s)f(s)dx \to \tfrac{1}{2}f(x) \text{ as } y \to 0$$

[24]Let me mention already here that this together with the circumstance that H^p and $\overline{H}^p$ are closed in L^p (by simple Banach space theory) implies (in fact is equivalent to) the existence of bounded projections onto H^p and $\overline{H}^p$.

[25]Formula L is equivalent to the more familiar

$$\int_R Q_\varepsilon(x-s)f(s)\,ds - \frac{1}{\pi}\int_{|x-s|>\varepsilon} \frac{f(s)}{x-s}\,ds \to 0;$$

but it also takes into account the fact that P_t is an approximate identity.

almost everywhere on R and in L^p norm, $1 \leqslant p < \infty$.

$$H_y f = \frac{2}{i} \int_{|x-s|>y} K_0(x-s) f(s) dx$$

is the (truncated) Hilbert transform. Because of (L), Theorem K (and Q) is equivalent to

THEOREM H: *For $f \in L^p$, $1 < p < \infty$, $\|H_y f\|_p \leqslant A_p \|f\|_p$, where A_p is a constant independent of f and y.*

There are famous real variable proofs of Theorem H, which then give new proofs of K and Q. It is not hard to show now that for y tending to zero $H_y f$ converges in L^p norm to a limit, which we denote by Hf, and that $\|Hf\|_p \leqslant A_p \|f\|_p$ with the same constant A_p as in Theorem H. The operator $f \mapsto Hf$ is the *Hilbert transform*. Via (L) this establishes the existence of the limit[26]

$$\lim_{y\to 0} K_y * f$$

and we have

$$\lim_{y\to 0} K_y * f = \tfrac{1}{2}(f + iHf).$$

More explicitly defining $Pf = \lim_{y\to 0} K_y * f$, we have

$$Pf(x) = \tfrac{1}{2} f(x) + \lim_{y\to 0} \int_{|x-s|>y} K_0(x-s) f(s) dx.$$

If $f \in H^p \cap L^2 = H^p \cap H^2$—a dense subspace of H^p—then we know from the H^2 theory that $K_y * f = f_y$, and therefore that $K_y * f \to f$ in L^p, i.e., $Pf = f$. By the continuity of P this holds for all $f \in L^p$. On the other hand since $K_y * f$, and hence Pf for

[26]The existence of lim $K_y * f$ can also be proved easily by complex variable methods.

arbitrary f in L^p, is in H^p; we finally have that P is a bounded projection onto H^p. The existence of this projection was alluded to in the footnote 24 on page 291. Finally, let us remark that because of the special relation $K_0(x) = (i/2)Q_0(x)$ for $x \neq 0$, we have

(H-Q) $$Hf = \lim_{y \to 0} Q_y * f.$$

To get a clear view of what happens in higher dimensions, it is best to consider Theorems C and K. They behave very differently when one tries to extend them. While C always extends, K is known to extend only in exceptional cases. A reason for this is that C is a statement internal to $H^p(\check{S})$ whereas K involves all of $L^p(\check{S})$ and (because of its link with the projection P) the relation of $L^p(\check{S})$ to $H^p(\check{S})$. It is this relation which is modified in an essential way if one passes from one to several variables: We have already seen for $p = 2$ that in several variables $H^p(\check{S})$ and $\overline{H^p}(\check{S})$ do not exhaust $L^p(\check{S})$ or, equivalently, that not every real element of $L^p(\check{S})$ is the real part of an $H^p(\check{S})$ function. This, as we shall presently see, strongly influences the relation between the relevant kernels and brings new ones onto the scene. Also the singularities of the new kernels become worse. Before taking up these matters let me first report on the extensions of Theorem C. If D is a Siegel domain of either type or a bounded symmetric domain and $f = u + iv \in H^p$ with $1 < p < \infty$, then we can, as before, define the conjugate function map $u \mapsto v$. Note that again it is defined without explicit reference to any kernel. The statement C remains true in all cases. For bounded symmetric domains this was proved by Bochner [4] (who actually proved it for the more general circular domains that he considered in that paper). For Siegel domains of type I it is due to Stein and G. Weiss [80, p. 130]. For Siegel domains of type II it is in [53].

Let us now turn to looking at the kernels in several variables. The first observation one has to make is that the real part of the Szegö kernel never can be the Poisson kernel. If it were then P_t would belong to $H^2 \oplus \overline{H}^2$ together with all its left-translates, so $f * P_t$ would be 0 for every f which is orthogonal to $H^2 \oplus \overline{H}^2$. However, $f * P_t \mapsto f$. The best thing now is to look at a couple of examples. Let us take $\mathcal{H}^2$ and the half-plane equivalent to B^2, Siegel domains of type I and II respectively.

For $\mathcal{H}^2$,

$$K_y(x) = K_{y_1, y_2}(x_1, x_2) = \frac{1}{-2\pi i(x_1 + iy_1)} \cdot \frac{1}{-2\pi i(x_2 + iy_2)}$$

$$= \frac{1}{4\pi^2}\left(\frac{y_1}{x_1^2 + y_1^2}\frac{y_2}{x_2^2 + y_2^2} - \frac{x_1}{x_1^2 + y_1^2}\frac{x_2}{x_2^2 + y_2^2}\right)$$

$$+ \frac{i}{4\pi^2}\left(\frac{x_1}{x_1^2 + y_1^2}\frac{y_2}{x_2^2 + y_2^2} + \frac{y_1}{x_1^2 + y_1^2}\frac{x_2}{x_2^2 + y_2^2}\right).$$

For the half-plane equivalent to the ball B^2,

$$K_t(x, w) = \frac{1}{\pi^2}\frac{1}{(t + |w|^2 - ix)^2}$$

$$= \frac{1}{\pi^2}\left[\frac{(t + |w|^2)^2 - x^2}{((t + |w|^2)^2 + x^2)^2} + i\frac{2x(t + |w|^2)}{((t + |w|^2)^2 + x^2)^2}\right].$$

Let us begin by observing that $\operatorname{Re} K$ is no longer positive. Next, it is easy to check that it is not in L^1. Therefore the study of K_t as a convolution kernel no longer reduces to that of $\operatorname{Im} K_t$. In particular, the analogue of Q no longer would imply the analogue of K, nor would the extension of (H-Q) hold.[27] In these examples it is easy to check that $K_t \in L^p$ for $1 < p < \infty$, and that hence $(f * K_t)(x)$ is meaningful in every L^p, $1 \leqslant p < \infty$. When no explicit formulas are known, this is a problem too. At any rate since K_t is bounded and in L^2, it always is in L^p with $p \geqslant 2$. If one sets $y_1 = y_2 = 0$ in the first example or $t = 0$ in the second example, the real part of K_0 no longer vanishes. (On the other hand $\operatorname{Im} K_0$ vanishes in the first example.) So the kernel of the analogue of the Hilbert transform is K_0, not $\operatorname{Im} K_0$. All this goes to show that in

[27] I should mention here that, by calculating in H^2 where everything is well defined, it is easy to see that the conjugate function map $u \mapsto v$ is given (as in the case of $\mathcal{H}$) by $v_t = u * 2 \operatorname{Im} K_t$, $u \in H^2 + \overline{H}^2$.

several variables one has to study *all* of K. Observe now that the singularities of Re $K_{0,0}$ in the first example are on $x_1x_2 = 0$—that is on the boundary of $\Omega \cup (-\Omega)$. This is always the case when $\dim \Omega > 1$. Notice also that in the first example (considered for $p = 2$ where all the operators are defined) the formula defining P would assume a rather different form. It would contain an integral extended over a translate of $\Omega \cup (-\Omega)$, i.e., one no longer would have a principal value integral whose kernel is a function (on $\check{S}$). When $\dim \Omega > 1$ the whole problem of finding explicit expressions for P (always in H^2 where there are no existence problems) by means of boundary integrals is practically untouched. It is known, though, that Pf can be written as a convolution with a tempered distribution on $\check{S}$ [53].

Let us now see what can be said about extensions of Theorem K. For Siegel domains of type I whose cone is polyhedral—i.e., is the intersection of finitely many half-spaces of $\check{S}$—Theorem K holds. If $\Omega = \cap_{j=1}^{k} H_j$(where each H_j is a half-space) then its characteristic function χ_Ω is $\chi_\Omega = \Pi_{j=1}^{k} \chi_{H_j}$. Using the Fourier transform one sees that the theorem can be proved by repeated application of the one variable result. This result is widely known; I think it is due to A. P. Calderón. For Siegel domains of type I whose cones are not polyhedral, it is highly unlikely that K holds. The reason for this is C. Fefferman's theorem which states that in dimension $n \geqslant 2$, the unit ball is not a multiplier of L^p for $p \neq 2$ [17].

There is one more case in which A. Korányi and I have shown that K extends and an explicit boundary integral representation for P holds [51, 52]. This is the case of the half-plane equivalent to the unit ball (and also the case of the unit ball itself). A look at the Cauchy kernel in the second example above shows that it is much more wholesome looking than the one for $\mathcal{H}^2$, because K_0 is singular at the origin only. Define $K^\epsilon(x, w)$ by

$$K^\epsilon(x, w) = \begin{cases} \dfrac{2^{n-2}\Gamma(n)}{\pi^n} \dfrac{1}{\left(|w|^2 - ix\right)^n} & \text{if } \left(x^2 + |w|^4\right)^{1/2} > \epsilon, \\ 0 & \text{if } \left(x^2 + |w|^4\right)^{1/2} \leqslant \epsilon. \end{cases}$$

It can be shown that the following analogue of formula (L) holds: $f * K_\epsilon - f * K^\epsilon \to \frac{1}{2} f$ as $\epsilon \to 0$ in L^p norm for $1 \leqslant p < \infty$ and almost everywhere on $\check{S}$. Furthermore for $f \in L^p(\check{S})$, $1 < p < \infty$, $\| f * K^\epsilon \|_p \leqslant A_p \| f \|_p$ independently of ϵ (which extends Theorem H), and $f * K^\epsilon$ converges in L^p to a limit Kf. These facts imply (the extension of) Theorem K and the representation of P in terms of a principal value integral on $\check{S}$.

The proof uses singular integrals on groups. The theory of these was initiated by A. Knapp and E. M. Stein in connection with the representation theory of semi-simple lie groups [37] [38]. The function $(x, w) \mapsto |(x, w)| = (|x|^2 + |w|^4)^{1/2}$ plays an important part in the proof.[28] Since I shall need it in discussing boundary behavior let me say a few words about it. Observe that it can be defined on the Šilov boundary of any Siegel domain of type II. (It generalizes the map $x \mapsto |x|$ and reduces to that map in the case of type I domains.) It is a generalized distance, and has the following interesting homogeneity property (which again generalizes the corresponding property of $|x|$ on Euclidean spaces): For $t > 0$ the mapping $\delta(t) : (x, w) \to (t^2 x, tw)$ is an automorphism of the group $\check{S}$, and $|\delta(t)(x, w)| = t^2 |(x, w)|$. (The reader is referred to the article of E. M. Stein in this volume where some aspects of this theory are discussed.)

We now come to the concluding section of this review. It deals with the convergence almost everywhere at the Šilov boundary of holomorphic functions, and, more generally, of Poisson integrals. It was mentioned earlier that Poisson integrals can be defined on all non-compact Riemannian symmetric spaces and that the more special theory which is our concern here developed simultaneously and in constant interaction with the more general one. Many of the results on Poisson integrals described below have—usually weaker—analogues in the more general context.[29]

The basic one variable result whose generalizations we consider

---

[28]This function is called a gauge in [52]. Knapp and Stein [38] call it a norm function. It is closely related to the notion of ergodic metric on a group which was introduced by Calderón [7].

[29]A. Korányi's article [47] surveys the whole subject in some detail.

here is Fatou's theorem [16; 91, I, p. 101]. It will be referred to as Theorem I (or just I) in the following. Its statement runs as follows: *Let U be the Poisson integral of a function $u \in L^p(T)$, $1 \leqslant p \leqslant \infty$. Then $U(z)$ tends to $u(z_0)$ for almost all $z_0 \in T$ as z approaches z_0 nontangentially.* Note that for $p > 1$ the theorem follows from the case $p = 1$ because $L^1(T) \supset L^p(T)$ for $p > 1$. An important corollary of Theorem I—actually of its special case $p = \infty$—is the following Theorem II: *Let $f \in H^p(U)$, $0 < p \leqslant \infty$.*[30] *Then f converges nontangentially almost everywhere on T, and the function f^* on T thus defined belongs to $H^p(T)$.* Both theorems remain true if U and T are replaced by $\mathcal{H}$ and R. Observe that II is new only for $p < 1$, and that for $1 < p < \infty$ (by virtue of M. Riesz's theorem on the conjugate function) it implies I. Generalizations of I to several variables will yield corresponding generalizations of II in the range $1 < p \leqslant \infty$ because holomorphic functions are the Poisson integrals of their boundary functions if these are in L^p, $1 < p \leqslant \infty$.[31] However, a generalization of Theorem II in the range $1 < p < \infty$ no longer states the same fact as the corresponding generalization of I because a real element of L^p, in general, is not the real part of an H^p function. Furthermore, speaking somewhat loosely, for $p < 1$ I does not imply II in several variables [73 p. 61]. Hence the latter has to be established independently.

The discussion of the generalizations of I and II must begin by giving a precise meaning to the term non-tangential convergence. Let us take Siegel domains of type I first. Let, then, T_Ω be such a domain. The point $z = x_0 + iy \in T_\Omega$ converges *unrestrictedly* to the point x_0 of the Šilov boundary of T_Ω if y tends to zero. (Recall y is in Ω.) Let now T_Ω be different from $\mathcal{H}$, and let Ω_0 be a proper open subcone of Ω (where vertex is also 0) such that $\overline{\Omega}_0 \subset \Omega \cup \{0\}$. The point $z = x_0 + iy \in T_\Omega$ tends *restrictedly* to x_0 if y tends

[30]The conclusion of theorem II remains true if f belongs to the Nevanlinna class N. It follows immediately from the fact that every $f \in N$ is of the form $f = g/h$ with g and $h \in H^\infty$ [15, p. 16].

[31]This fact is also true for $p = 1$; but for Siegel domains of type II the proof requires some form of (a generalization of) theorem II.

to zero, but stays in Ω_0. The point $z = x + iy \in T_\Omega$ tends *nontangentially and unrestrictedly* to $x_0 \in \check{S}$ if z tends to x (i.e., $x \to x_0$ and $y \to 0$), and $|x - x_0| < c|y|$ (eventually) for some c, $0 < c < 1$. If y stays in Ω_0 during this process, then the convergence is said to be *nontangential and restricted.*

Let now $D = D_{\Omega, \Phi}$ be a Siegel domain of type II. Recall that its Šilov boundary is $\check{S} = \{(z, w) | \operatorname{Im} z - \Phi(w, w) = 0\}$, and that a point in D can be written as $(x + i(t + \Phi(w, w)), w)$ with $t \in \Omega$. The point $(x_0 + i(t + \Phi(w_0, w_0), w_0)$ of D tends unrestrictedly to the point $(x_0 + i\Phi(w_0, w_0), w_0)$ if t tends to zero. If Ω_0 is as before and y (eventually) stays in Ω_0, then the convergence is called restricted. One could define nontangential convergence, but there is a more natural mode of convergence to the Šilov boundary for these domains, which was introduced by Korányi in [45]—admissible convergence. It is more "natural" because it has group invariance properties that nontangential convergence does not have. (For domains of type I, nontangential convergence is the "natural" notion.) Recall that on S there is a gauge $(x, w) \to \sqrt{x^2 + |w|^4}$. Let $\alpha > 0$, then the set $\Gamma_\alpha(0, 0) = \{(x + it, w) \in D | \sqrt{x^2 + |w|^4} < \alpha|t|\}$ is called an *admissible domain* at the origin $(0, 0)$. If $\dim \Omega > 1$[32] and if Ω_0 is a proper open subcone of Ω such that $\bar{\Omega}_0 \subset \Omega \cup \{0\}$ then

$$\Gamma_{\alpha, \Omega_0}(0, 0) = \left\{(x + it, w) \in D | t \in \Omega_0, \sqrt{x^2 + |w|^4} < \alpha|t|\right\}$$

is a *restricted admissible domain* at the origin. The corresponding domains $\Gamma_\alpha(x_0, w_0)$ and $\Gamma_{\alpha, \Omega_0}(x_0, w_0)$ at the point (x_0, w_0) of the Šilov boundary are obtained by translating $\Gamma_\alpha(0, 0)$ and $\Gamma_{\alpha, \Omega_0}(0, 0)$ to (x_0, w_0). A point $(x + it, w)$ of D converges admissibly (admissibly and restrictedly) to the point (x_0, w_0) of $\check{S}$ if it (eventually) stays in $\Gamma_\alpha(x_0, w_0)$ ($\Gamma_{\alpha, \Omega_0}(x_0, w_0)$). To illustrate how this notion

---

[32]It is easy to see that in each dimension there is only one domain with $\dim \Omega = 1$—the one equivalent to the ball. For these domains the distinctions between (unrestricted) admissible and restricted admissible convergence disappears.

generalizes that of nontangential convergence let us consider the simplest case: Let $D = \{(x + it, w) | x, t \in R, w \in C, t > 0\}$ be the half-plane equivalent to B^2, so that $\Gamma_1(0, 0) = \{(x + it, w) | x \in R, w \in C, t > 0, (x^2 + |w|^4)^{1/2} < t\}$. Let us look at the intersection of $\Gamma_1(0, 0)$ with the complex subspace $w = 0$ of C^2:

$$\Gamma_1(0, 0) \cap \{w = 0\} = \{(x + it, 0) | x \in R, t > 0, |x| < t\}.$$

This is just a usual non-tangential domain in the upper half-plane at 0. Therefore, restricted to the plane $w = 0$ admissible convergence is ordinary non-tangential convergence. If we intersect $\Gamma_1(0, 0)$ with the real subspace $x = 0$ we get $\Gamma_1(0, 0) \cap \{x = 0\} = \{(it, w) | w \in \mathbf{C}, |w|^2 < t\}$. Figure 2 shows that restricted to $\{x = 0\}$, admissible convergence may become tangential.

Let us now see what the generalizations of I and II are. I begin with II. Let $D = U^n$ or $\mathcal{H}^n$, $\check{S} = T^n$ or R^n respectively. If $f \in H^p$, $0 < p \leqslant \infty$, then f converges non-tangentially and unrestrictedly to f^* almost everywhere on $\check{S}$. For U^n this result is due to A. Zygmund [93; 91, II, p. 318]. For $\mathcal{H}^n$ it was proved by E. M. Stein, G. Weiss, and M. Weiss in [79]. This paper also contains the following extension of II: Let T_Ω be a Siegel domain of type I different from $\mathcal{H}^n$. Let $f \in H^p(T_\Omega)$, $0 < p \leqslant \infty$. Then f converges non-tangentially and restrictedly almost everywhere on $\check{S}$. This is the first result which carries the theory of almost every-

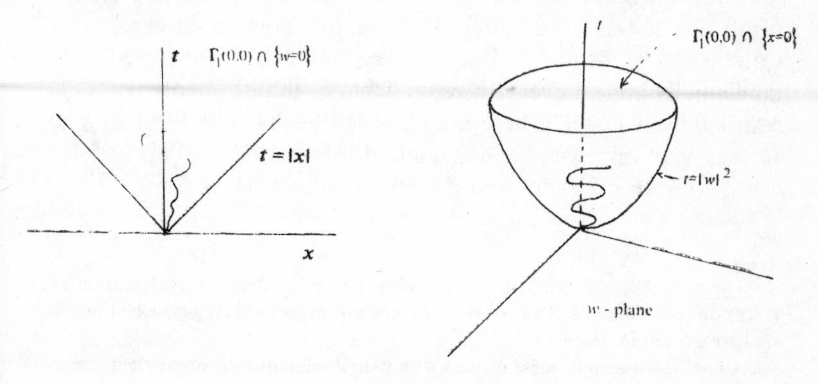

FIG. 2

where convergence well beyond the traditional classical setting.[33] The analogous theorem for Siegel domains of type II was proved by Stein, but with the weaker conclusion that f converges to f^* restrictedly [76]. Unless further conditions are imposed (D symmetric and $1 \leqslant p \leqslant \infty$, or D arbitrary but $p = \infty$) in which case —as we shall see presently—stronger results are obtained from extensions of Theorem I, this is so far the best result for domains of type II. The question of whether restricted convergence can be improved to restricted admissible convergence constitutes an interesting open problem.

In the present context the unit ball again plays its exceptional role. Zygmund proved that if $f \in N(B)$ then f converges non-tangentially to f^* almost everywhere on $\check{S} = \partial B$ [92]. This result was improved by Korányi who showed that non-tangential convergence can be replaced by admissible convergence [46]. There is a similar result for the polydisc due to Calderón and Zygmund [8] who showed that for $f \in N(U^n)$, f converges non-tangentially and restrictedly almost everywhere on T^n to f^*. No results of this type involving N seem to be known for more general domains.

Fatou's theorem stands in close connection with the differentiation of integrals, and its extensions to higher dimensions depend on suitable differentiation theorems which in turn depend on estimates for generalizations of the Hardy-Littlewood maximal function.[34]

In surveying the extensions of Theorem I let me again state the result for the polydisc first. If U is the Poisson integral of a function $u \in L^p(T^n)$, $1 \leqslant p \leqslant \infty$, then U converges non-tangentially almost everywhere on T^n to u: unrestrictedly if $1 < p$, restrictedly if $p = 1$. These are classical results: due for $1 < p$ to Jessen, Marcinkiewicz, and Zygmund [34; 91, II, p. 308]; and for $p = 1$ to Marcinkiewicz and Zygmund [58; 91, II, p. 309].

---

[33]The book [80] of E. M. Stein and G. Weiss contains a detailed treatment of H^p spaces on tube domains. Some of the results whose proofs are only sketched in [79] are here proved in detail.

[34]A brief discussion of these matters with special reference to symmetric domains can be found in Stein's article [77].

For Siegel domains which are half-plane realizations of bounded symmetric domains we have the following extension of I: Let D be a symmetric Siegel domain (of type I or II) and let $u \in L^p(\check{S})$ with $1 \leqslant p \leqslant \infty$. Then the Poisson integral of u converges almost everywhere on $\check{S}$ to u: restrictedly and non-tangentially if D is of type I, restrictedly and admissibly if D is of type II. For domains of type I and $1 < p$ the result is due to N. J. Weiss [85, 86], for the half-plane equivalent to the ball and in the full range $1 \leqslant p \leqslant \infty$ to Korányi [45].[35] For all other domains of type II and $p > 1$, it was proved by N. J. Weiss. The L^1 theorem is the joint work of Stein and N. J. Weiss [78]. For arbitrary Siegel domains and $p = \infty$, the analogous result was proved by A. Korányi and E. M. Stein [50].

There is an interesting negative result of Stein and N. J. Weiss [78], originally discovered in a special case in [79], which has no analogue in the classical cases of U^n and $\mathcal{H}^n$. Let D be the half-plane realization of an irreducible bounded symmetric space different from the ball. Then there exists a function $u \in L^\infty(\check{S})$ whose Poisson integral almost everywhere on $\check{S}$ fails to converge unrestrictedly to u. Furthermore, for every $1 \leqslant p \leqslant \infty$ there exists a $u \in L^p(\check{S})$ such that its Poisson integral U has the property that for almost all $z_0 \in \check{S}$ lim sup $U(z) = +\infty$ as z tends to z_0 unrestrictedly.[36] For $p \neq \infty$ surprisingly little is known about the Poisson integrals for Siegel domains which are not symmetric, even if the domains are of type I or homogeneous. The main reason for this is that the estimates used in proving the extensions of I for $p \neq \infty$ rely on explicit formulas for the Poisson kernels. Such formulas for general domains are not available. N. J. Weiss points out in [85] that the results of that paper (I for $1 < p$) can be proved for the four dimensional homogeneous non-symmetric domain which occurs in Pjateckiĭ-Šapiro's solution of Cartan's

---

[35]Recall that in this case the distinction between restricted admissible and unrestricted admissible convergence disappears.

[36]The paper [87] of N. J. Weiss has a simplified proof of these facts. This paper also contains simplified proofs of the results of [85], [86] and [78] for two special domains.

problem. The case of T_Ω where the "cross section" of Ω is a cube in R^3 was studied by B. S. Kabak [35], and domains of type I with a larger class of cones by L. J. Dickson [13].

I have not mentioned any results on the boundary behavior almost everywhere of Poisson integrals for the bounded symmetric domains other than U^n and B^n. Results of this nature can be derived from the corresponding ones for half-spaces [47].

A SHORT GUIDE TO FURTHER READING

SYMMETRIC SPACES

There is an excellent, very readable, expository survey of symmetric space theory, Lie groups, Lie algebras, etc., by Helgason [27]. A very clear and short introduction to symmetric spaces can be found in chapter VIII of Wolf's book [89]. The subject is treated in full detail in the textbooks of Helgason [26] and Loos [57]. The former has a chapter on the bounded symmetric domains.

CARTAN AND SIEGEL DOMAINS

The necessary background from the theory of several complex variables is found in Narasimhan's book [62]. The classical texts on the domains—they do not use the general theory of symmetric spaces—are the books of Hua [28] and Pjateckiĭ-Šapiro [67]. The former also contains much material about the kernel functions and harmonic analysis. A short introduction which does use some symmetric space theory, discusses the general Cayley transforms and also some function theory is found in Korányi's C.I.M.E. notes [43]. The boundary structure of the Cartan domains is treated in great detail by Wolf in [90]. Murakami's Springer notes [61] treat homogeneous Siegel domains in detail. Koecher's Jordan algebra approach to the theory of Cartan domains is in [41].

FUNCTION THEORY

H^p theory in one variable is treated in detail in Duren's book [15]. A short introduction to the subject is in chapter VII of Zygmund's treatise [91]. The function theory of the polydisc is treated in detail in Rudin's monograph [73]. The theory of H^p spaces on tube domains is treated in chapter III of the book [80] of Stein and Weiss. The Springer notes [12] of Coifman and Weiss have three very readable chapters on function theory and harmonic analysis on the two dimensional unit ball. Hua's book [28] treats harmonic analysis on the classical domains, but it was written in the fifties and therefore does not contain the newer results. The basic facts about H^2 theory and the Poisson integral for Siegel domains of type II is outlined in Korányi's notes [43]. The role of Cartan domains in the theory of automorphic functions is discussed in Baily's book [3], which also contains some material concerning analysis on these domains. There is no comprehensive treatment of the subject in the literature.

REFERENCES

1. Alexander, H., "Holomorphic mappings from the ball and polydisc," *Math. Ann.*, **209** (1974), 249–256.
2. Auslander, L. and R. Tolimieri, *Abelian harmonic analysis, theta functions and function algebras on a nilmanifold*, Lecture Notes in Mathematics 436, Springer, New York, 1975.
3. Baily, W. L., *Introductory Lectures on Automorphic Forms*, Princeton University Press, Princeton, 1973.
4. Bochner, S., "Classes of holomorphic functions of several variables in circular domains," *Proc. Nat. Acad. Sci. U.S.A.*, **46** (1960), 721–723.
5. ——, "Group invariance of Cauchy's formula in several variables," *Ann. of Math.*, **45** (1944), 686–707.
6. Borel, A., "Kaehlerian coset spaces of semi-simple Lie groups," *Proc. Nat. Acad. Sci. U.S.A.*, **40** (1954), 1147–1151.
7. Calderón, A. P., "A general ergodic theorem," *Ann. of Math.*, **58** (1953), 182–191.
8. ——, and A. Zygmund, "Note on the boundary values of functions of several variables," in *Contributions to Fourier Analysis, Ann. of Math. Studies No. 25*, Princeton, 1950.
9. Cartan, É., "Sur les domaines bornés homogènes de l'espace de n variables complexes," *Abh. Math. Sem. Univ. Hamburg*, **11** (1935), 116–162.

10. Cartan, H., *Sur les groupes de transformations analytiques*, Actualités Sci. Indust., Exposés Math. IX, Hermann, Paris, 1935.

11. Cartier, P., "Quantum mechanical commutation relations and theta functions," in *Proc. Symposia in Pure Math. IX*, Amer. Math. Soc., Providence, 1965.

12. Coifman, R. R., and G. Weiss, *Analyse harmonique non-commutative sur certains espaces homogènes*, Lecture Notes in Mathematics 242, Springer, New York, 1971.

13. Dickson, L. J., "Limit properties of Poisson kernels of tube domains," *Trans. Amer. Math. Soc.*, **182** (1973), 383–401.

14. Dorfmeister, J., *Eine Theorie der konvexen regulären Kegel*, Thesis, University of Münster, 1974.

15. Duren, P., *Theory of H^p Spaces*, Academic Press, New York, 1971.

16. Fatou, P., "Séries trigonométriques et séries de Taylor," *Acta Math.*, **30** (1906), 335–400.

17. Fefferman, C., "The multiplier problem for the ball," *Ann. of Math.*, **94** (1971), 330–336.

18. Folland, G. B., and E. M. Stein, "Estimates for the $\bar{\partial}_b$ complex and analysis on the Heisenberg group", *Comm. Pure Appl. Math.*, **27** (1974), 429–522.

19. Furstenberg, H., "A Poisson formula for semi-simple Lie groups," *Ann. of Math.*, **77** (1963), 335–386.

20. ——, "Boundaries of Riemannian symmetric spaces," in *Symmetric Spaces*. Short courses presented at Washington University, W. M. Boothby and G. L. Weiss, editors, Marcel Dekker, Inc., New York, 1972.

21. Gindikin, S. G., "Analysis on homogeneous domains," *Uspehi Mat. Nauk*, **19** (1964), 3–92 (Russian). Translated in *Russian Math. Surveys*, **19** (1964), 1–89.

22. Godement, R., "Une généralisation du théorème de la moyenne pour les fonctions harmoniques," *C. R. Acad. Sci. Paris*, **234** (1952), 2137–2139.

23. Hahn, K. T., and J. Mitchell, "H^p spaces on bounded symmetric domains," *Trans. Amer. Math. Soc.*, **146** (1969), 521–531.

24. ——, "H^p spaces on bounded symmetric domains," *Ann. Polon. Math.*, **28** (1973), 89–95.

25. Hano, J. I., "On Kählerian homogeneous spaces of unimodular Lie groups," *Amer. J. Math.*, **79** (1957), 885–900.

26. Helgason, S., *Differential Geometry and Symmetric Spaces*, Academic Press, New York, 1962.

27. ——, "Lie groups and symmetric spaces," in *Battelle Rencontres 1967*, Benjamin, New York, 1968.

28. Hua, L. K., "Harmonic Analysis of Functions of Several Complex Variables in the Classical Domains," *Trans. Math. Monographs 6*, Amer. Math. Soc., Providence, 1963.

29. ——, "On the theory of automorphic functions of a matrix variable, I. Geometric Basis," *Amer. J. Math.*, **66** (1944), 470–488.

30. Igusa, J., *Theta Functions*, Springer, New York, 1972.

31. Ise, M., "On canonical realizations of bounded symmetric domains as matrix spaces," *Nagoya Math. J.*, **42** (1971), 115–133.

32. ——, "Realization of irreducible bounded domains of Type (V)," *Proc. Japan Acad.*, **45** (1969), 233–237.

33. ——, "Realization of irreducible bounded domains of Type (VI)," *Proc. Japan Acad.*, **45** (1969), 846–849.

34. Jessen, B., J. Marcinkiewicz and A. Zygmund, "Note on the differentiability of multiple integrals," *Fund. Math.*, **25** (1935), 217–34.

35. Kabak, B. S., Thesis, Yeshiva University, 1970.

36. Kaup, W., "Reelle Transformationsgruppen und invariante Metriken auf komplexen Räumen," *Invent. Math.*, **3** (1967), 43–70.

37. Knapp, A. W., and E. M. Stein, "Singular integrals and the principal series," *Proc. Nat. Acad. Sci. U.S.A.*, **63** (1969), 281–284, ibid., **66** (1970), 13–17.

38. ——, "Intertwining operators for semi-simple Lie Groups," *Ann. of Math.*, **93** (1971), 489–578.

39. Koecher, M., "Positivitätsbereiche im R^n," *Amer. J. Math.*, **79** (1957), 575–596.

40. ——, "Jordan algebras and their applications," mimeographed notes, University of Minnesota, 1962.

41. ——, *An elementary approach to bounded symmetric domains*, Rice University, Houston, 1969.

42. Korányi, A., "A Poisson integral for homogeneous wedge domains," *J. Analyse Math.*, **14** (1965), 275–284.

43. ——, "Holomorphic and harmonic functions on bounded symmetric domains," C.I.M.E. Summer course on the Geometry of Bounded Homogeneous Domains, Cremonese, Roma, 1968, 125–197.

44. ——, "The Poisson integral for generalized half-planes and bounded symmetric domains," *Ann. of Math.*, **82** (1965), 332–350.

45. ——, "Harmonic functions on hyperbolic hermitian space," *Trans. Amer. Math. Soc.*, **135** (1969), 507–516.

46. ——, "A remark on boundary values of functions of several variables," in *Several Complex Variables I, Maryland 1970,* Proceedings of the International Mathematical Conference, held at College Park, April 6–17, 1970. Lecture Notes in Mathematics 155, Springer, New York, 1970.

47. ——, "Harmonic functions on symmetric spaces," in *Symmetric Spaces*. Short courses presented at Washington University, W. M. Boothby and G. L. Weiss, editors, Marcel Dekker, Inc., New York, 1972.

48. ——, "H^p spaces on compact nilmanifolds," *Acta Sci. Math. (Szeged)*, **34** (1973), 175–190.

49. Korányi, A., and E. M. Stein, "H^2 spaces of generalized half-spaces," *Studia Math.* **44** (1972), 379–388.

50. ——, "Fatou's theorem for generalized half-planes," *Ann. Scuola Norm. Sup. Pisa*, **22** (1968), 107–112.

51. Korányi, A., and S. Vági, "Intégrales singulières sur certains espaces homogènes," *C. R. Acad. Sci. Paris*, **268** (1969), 765–768.

52. ——, "Singular integrals on homogeneous spaces and some problems of classical analysis," *Ann. Scuola Norm. Sup. Pisa*, **25** (1971), 575–648.

53. Korányi, A., S. Vági, and G. V. Welland, "Remarks on the Cauchy integral and the conjugate function in generalized half-planes," *J. Math. Mech.*, **19** (1970), 1069–1081.

54. Korányi, A., and J. A. Wolf, "The realization of hermitian symmetric spaces as generalized half-planes," *Ann. of Math.*, **81** (1965), 265–288.

55. Koszul, J. L., "Sur la forme hermitienne canonique des espaces homogènes complexes," *Canad. J. Math.*, **7** (1955), 562–576.

56. Lowdenslager, D., "Potential theory in bounded homogeneous domains," *Ann. of Math.*, **67** (1958), 467–484.

57. Loos, O., *Symmetric Spaces* I, II, Benjamin, New York, 1969.

58. Marcinkiewicz, J., and A. Zygmund, "On the summability of double Fourier series," *Fund. Math.*, **32** (1939), 112–132.

59. Mitchell, J., "Potential theory in the geometry of matrices," *Trans. Amer. Math. Soc.*, **79** (1955), 401–422.

60. Moore, C. C., "Compactifications of symmetric spaces II: the Cartan domains," *Amer. J. Math.*, **86** (1964), 358–378.

61. Murakami, S., *On automorphisms of Siegel domains*, Lecture Notes in Mathematics 286, Springer, New York, 1972.

62. Narasimhan, R., *Several Complex Variables*, University of Chicago Press, Chicago, 1971.

63. Neumann, J. von, "Die Eindeutigkeit der Schrödingerschen Operatoren," *Math. Ann.*, **104** (1931), 570–578.

64. Ogden, R. D., and S. Vági, "Harmonic analysis and H^2 functions on Siegel domains of type II," *Proc. Nat. Acad. Sci. U.S.A.*, **60** (1972), 11–14.

65. Pjateckiĭ-Šapiro, I. I., "On a problem proposed by É. Cartan," *Dokl. Akad. Nauk SSSR*, **124** (1959), 272–273 (Russian).

66. ——, "The geometry of homogeneous domains in the theory of automorphic functions." Solution of a problem of É. Cartan, *Uspehi Mat. Nauk*, **14** (1959), no. 3, 190–192.

67. ——, *Geometry of classical domains and theory of automorphic functions*, Fizmatgiz, Moscow, (1961) (Russian). French translation, Dunod, Paris, 1966. English translation, Gordon and Breach, New York, 1969.

68. Poincaré, H., "Les fonctions analytiques de deux variables et la représentation conforme," *Rend. Circ. Mat. Palermo*, **23** (1907), 185–220.

69. Pukánszky, L., *Leçons sur les représentations des groupes*, Dunod, Paris, 1967.

70. Riesz, M., "Sur les fonctions conjuguées," *Math. Zeit.*, **27** (1927), 218–244.

71. Rothaus, O. S., "Domains of positivity," *Abh. Math. Sem. Univ. Hamburg*, **24** (1960), 189–235.

72. ——, "The construction of homogeneous cones," *Ann. of Math.*, **83** (1966), 358–376.

73. Rudin, W., *Function Theory in Polydiscs*, Benjamin, New York, 1969.

74. Schmid, W., "Die Randwerte holomorpher Funktionen auf hermitesch symmetrischen Räumen," *Invent. Math.*, **9** (1969), 61–80.

75. Siegel, C. L., "Symplectic geometry," *Amer. J. Math.*, **63** (1943), 1–86.

76. Stein, E. M., "Note on the boundary values of holomorphic functions," *Ann. of Math.*, **82** (1965), 351–353.

77. ——, "The analogues of Fatou's theorem and estimates for maximal functions," C.I.M.E. summer course on the Geometry of Bounded Homogeneous Domains, Cremonese, Roma, 1968, 289–307.

78. ——, and N. J. Weiss, "On the convergence of Poisson integrals," *Trans. Amer. Math. Soc.*, **140** (1969), 35–54.

79. ——, G. Weiss, and M. Weiss, "H^p classes of holomorphic functions in tube domains," *Proc. Nat. Acad. Sci. U.S.A.*, **52** (1964), 1035–1039.

80. ——, and G. Weiss, *Introduction to Fourier Analysis on Euclidean Spaces*, Princeton University Press, Princeton, 1971.

81. Stoll, M., "Harmonic majorants for plurisubharmonic functions on bounded symmetric domains with applications to the spaces H_Φ and N_*," to appear in *J. Reine Angew. Math.*

82. Vági, S., "On the boundary values of holomorphic functions," *Rev. Un. Mat. Argentina*, **25** (1970), 123–136.

83. Vinberg, E. B., "Theory of convex homogeneous cones," *Trudy Moskov. Mat. Obšč.*, **12** (1963), 303–358 (Russian). Translated in *Transactions of the Moscow Mathematical Society for the year 1963*, Amer. Math. Soc., Providence, 1965.

84. ——, S. G. Gindikin and I. I. Pjateckiĭ-Šapiro, "Classification and canonical realization of complex bounded homogeneous domains," *Trudy Moskov. Mat. Obšč.*, **12** (1963), 359–388. Translated in *Transactions of the Moscow Mathematical Society for the year 1963*, Amer. Math. Soc., Providence, 1965.

85. Weiss, N. J., "Almost everywhere convergence of Poisson integrals on tube domains over cones," *Trans. Amer. Math. Soc.*, **129** (1967), 283–307.

86. ——, "Convergence of Poisson integrals on generalized upper half-planes," *Trans. Amer. Math. Soc.*, **136** (1969), 109–123.

87. ——, "Fatou's theorem for symmetric spaces," in *Symmetric Spaces*, W. M. Boothby and G. L. Weiss, editors, Marcel Dekker, Inc., New York, 1972, 413–441.

88. Weyl, H., *Classical Groups*, Princeton University Press, Princeton, 1939.

89. Wolf, J. A., *Spaces of Constant Curvature*, McGraw-Hill, New York, 1967.

90. ——, "Fine structure of hermitian symmetric spaces," in *Symmetric Spaces*. Short courses presented at Washington University, W. M. Boothby and G. L. Weiss, editors, Marcel Dekker, Inc., New York, 1972.

91. Zygmund, A., *Trigonometric Series I, II,* Cambridge University Press, Cambridge, 1959.

92. ——, "A remark on functions of several complex variables," *Acta Sci. Math. (Szeged)*, **12** (1950), 66–68.

93. ——, "On the boundary values of functions of several complex variables," *Fund. Math.*, **36** (1949), 207–235.

INDEX